Basic VLSI Design Technology

Technical Questions and Solutions

RIVER PUBLISHERS SERIES IN CIRCUITS AND SYSTEMS

Indexing: All books published in this series are submitted to the Web of Science Book Citation Index (BkCI), to SCOPUS, to CrossRef and to Google Scholar for evaluation and indexing.

The "River Publishers Series in Circuits and Systems" is a series of comprehensive academic and professional books which focus on theory and applications of Circuit and Systems. This includes analog and digital integrated circuits, memory technologies, system-on-chip and processor design. The series also includes books on electronic design automation and design methodology, as well as computer aided design tools.

Books published in the series include research monographs, edited volumes, handbooks and textbooks. The books provide professionals, researchers, educators, and advanced students in the field with an invaluable insight into the latest research and developments.

Topics covered in the series include, but are by no means restricted to the following:

- Analog Integrated Circuits
- Digital Integrated Circuits
- Data Converters
- Processor Architecures
- System-on-Chip
- Memory Design
- Electronic Design Automation

For a list of other books in this series, visit www.riverpublishers.com

Basic VLSI Design Technology

Technical Questions and Solutions

Cherry Bhargava

Lovely Professional University
India

Gaurav Mani Khanal

University of Rome Tor Vergata
Italy

Published 2020 by River Publishers
River Publishers
Alsbjergvej 10, 9260 Gistrup, Denmark
www.riverpublishers.com

Distributed exclusively by Routledge
4 Park Square, Milton Park, Abingdon, Oxon OX14 4RN
605 Third Avenue, New York, NY 10017, USA

Basic VLSI Design Technology Technical Questions and Solutions / by Cherry Bhargava, Gaurav Mani Khanal.

Routledge is an imprint of the Taylor & Francis Group, an informa business

ISBN 978-87-7022-158-0 (print)

This Book is dedicated to

My Parents, Husband
and
Loving Daughters Mishty & Mauli

— Cherry Bhargava

My Parents, Wife
and Daughter Gaurisha

— Gaurav Mani Khanal

Contents

Preface

With the recent technological advancement and rapid development of compact devices, the world is accelerating toward the concept of "integration." The integration of millions of transistors on a single chip has led to the era of very large scale integration (VLSI). It is an interdisciplinary science of utilizing advance semiconductor technology to create various functions of computer systems. The emergence of VLSI technology has opened many new architectural concepts that depart sharply from the design of conventional computer systems. The design of single-chip computers, special-purpose chips to implement algorithms, and more intelligent computer system modules, such as memory and I/O processors, are the basic impacts of VLSI technology on computer system design.

VLSI is all about integrated circuit (IC) design using VLSI at the nanometer level. All the ICs that we use in mobiles, television sets, computers, satellites, automobiles, etc., are designed by VLSI technology. Today various companies like Texas Instruments, Infineon, Alliance Semiconductors, Cadence, Synopsys, Cisco, Micron Tech, National Semiconductors, ST Microelectronics, Qualcomm, Lucent, Mentor Graphics, Analog Devices, Intel, Philips, Motorola, and many other firms have proved themselves and are dedicated to many fields of VLSI such as programmable logic devices, hardware descriptive languages, design tools, etc. The purpose of writing this book is to demystify the hardware design for software engineers as well as to elucidate the VLSI design and related technologies for students as well as researchers. This book refreshes the basic concepts for designing and developing an IC using various tools and techniques.

Organization of the Book

This book is divided into 5 units, which cover all the subdomains of basic VLSI design. After summarizing the basic concepts of related technology, the technical questions related to that particular technology is given with solutions.

Unit 1: Gives an introduction to VLSI technology. The VLSI design technologies such as combinational and sequential logic design techniques are described briefly. The various design issues and related questions are discussed at the end.

Unit 2: Explains the various steps involved in designing an IC. The different techniques required from transforming sand to silicon are depicted in brief.

Unit 3: Gives an overview of the complementary metal–oxide–semiconductor (CMOS) and the designs based on CMOS. The low power CMOS design and noise issues are described in short.

Unit 4: Focuses on the validation of static timing performance of a design by introducing various timing paths and violations. The concept of clocking and metastability is discussed in brief along with interview questions.

Unit 5: Illustrates the advancement in the field of semiconductors as well as the design process. Future trends in VLSI technology are discussed in this chapter.

Annexure I: List of Digital Circuit IC Numbers

Annexure II: List of Keywords, System Tasks, and Compiler Directives used in Verilog HDL

Acknowledgment

At this movement of our substantial enhancement, before we get into the thick of the things, we would like to add a few heartfelt words for the people who gave their unending support with their unfair humor and warm wishes. First and foremost, praises and thanks to the God, the Almighty, for his showers of blessings throughout, to complete this book successfully.

We want to acknowledge our students who provided us with the impetus to write a more suitable text. Our supporting family members deserve great acknowledgement in true sense who have always been a force to keep me riveted to our dedication towards the present book. Besides, we thank all our friends, well-wishers, respondents and academicians who helped throughout my journey from inception to completion.

List of Figures

List of Tables

List of Abbreviations

ABEL	Advanced Boolean Expression Language
ADC	Analog to Digital Converter
AHDL	Altera Hardware Description Language
ALU	Arithmetic Logic Unit
AM	Amplitude Modulation
ASIC	Application Specific Integrated Circuit
ATPG	Automatic Test Pattern Generation
BiCMOS	Bipolar Complementary Metal-Oxide Semiconductor logic
BIT	Binary Digit
BJT	Bipolar Junction Transistor
CB	Common Base
CC	Common Collector
CDMA	Code Division Multiple Access
CE	Common Emitter
CISC	Complex Instruction Set Computer
CMOS	Complementary Metal Oxide Semiconductor
CMRR	Common Mode Rejection Ratio
CNTFET	Carbon Nanotube Field-Effect Transistors
CRO	Cathode Ray Oscilloscope
CUPL	Compiler for Universal Programming Language
CVD	Chemical Vapor Deposition
DAC	Digital to Analog Converter
DRC	Design Rule Check
DTL	Diode Transistor Logic
ECL	Emitter Coupled Logic
EMF	Electro Motive Force
FA	Full Adder
FACT	Fairchild Advanced CMOS Technology
FDMA	Frequency Division Multiple Access
FET	Field Effect Transistor
FF	Flip Flop

FIT	Failure In Time
FM	Frequency Modulation
FPGA	Field Programmable Gate Array
GSI	Giga Scale Integration
HA	Half Adder
HDL	Hardware description language
I^2L	Integrated Injection Logic
IC	Integrated Circuit
IGFET	Insulated Gate field-effect transistor
ISUB	Subthreshold (weak inversion) Leakage
JFET	junction field-effect transistor
K-Map	Karnaugh Map
LSB	Least Significant Bit
LSI	Large Scale Integration
LVS	Layout vs Schematic
MOS	Metal Oxide Semiconductor
MSB	Most Significant Bit
MSI	Medium Scale Integration
MTBF	Mean Time Between Failure
MTTF	Mean Time To Failure
MTTR	Mean Time To Repair
PAL	Programmable Array Logic
PCB	Printed Circuit Board
Pd	Propagation Delay
PHDL	Printed Circuit Board HDL
PIPO	Parallel In Parallel Out
PISO	Parallel In Serial Out
PLA	Programmable Logic Array
PLD	Programmable Logic Devices
PTL	Pass-transistor logic
PUD	Pull Down Network
PUN	Pull Up Network
PV	Photo Voltaic
Pw	Pulse Width
QCA	Quantum-dot Cellular Automata
RAM	Random Access Memory
RF	Radio Frequency
RISC	Reduced Instruction Set Computer
ROM	Read Only Memory

RTL	Resistor Transistor Logic
RTL	Register Transfer Level
SAM	Sequential Access Memory
SCCMOS	Super Cut-off CMOS Technique
SET	Single Electron Transistor
SIPO	Serial In Parallel Out
SISO	Serial In Serial Out
SOC	System On Chip
SOI	Silicon On Insulator
SSI	Small Scale Integration
Tf	Falling time
Th	Hold time
Tr	Rising time
Ts	Settling time
TTL	Transistor-Transistor Logic
ULSI	Ultra Large Scale Integration
VHDL	Very high-speed integrated circuit Hardware Description Language
VLSI	Very Large Scale Integration
Vsb	Voltage between Source and Bulk(substrate)
VTCMOS	Variable Threshold CMOS

1

Digital System Design

1.1 Introduction

In the modern age, semiconductor electronics are advancing at an accelerated pace. Consequently, there is a major change in hardware design methodologies. The cost and space constraints evolve the integration of millions of components on a single chip. In the 1960s and 1970s, designers struggled with bulky vacuum tubes and other electronic components for optimization. Developing of a printed circuit board was a herculean task in past years. The traditional approach was to choose the components as per specification; decide the hardware and software specification; develop a printed circuit board (PCB); and test the entire system.

In the past three decades, the hardware used to be very bulky due to discrete components used in circuits. Subsequently, many integrated circuits (ICs) that can carry out specific functions in an inefficient manner made life a bit easier. The designer still strives to reduce the number of components. The higher the component count, the higher is the power consumption and the higher will be the failure rate. So, reliability becomes a challenging issue while designing millions of transistors or other components on a single chip. The development cost, time, and production cost will be high. Due to these drawbacks, the designer and hardware engineers are no longer using the traditional approach of having a large number of components on a PCB.

Now the objective is to develop a single IC which can perform which will achieve the entire system functionality at reduced cost and small size. This has evolved the technology of integration. Very large scale integration (VLSI) is the process of creating an IC by combining thousands of transistors into a single chip. An excellent example of VLSI technology is a mobile phone, which is sleek in shape, low power consuming, and portable to use. With the evolution of fast processors and EDA tools, designing and developing an IC becomes an interesting and easy task.

Table 1.1 IC integration level

Level of Integration	Number of Gates/Chips
Small-scale integration (SSI)	<12
Medium-scale integration (MSI)	12–99
Large-scale integration (LSI)	1000
Very large scale integration (VLSI)	10 k
Ultra large scale integration (ULSI)	100 k
Giga-scale integration (GSI)	1 Meg

The fast pace of developments in semiconductor technology has led to the development of small chips that can hold a lot of logic and memory. The examples of such technologies are field programmable gate array (FPGA), application specific integrated circuit (ASIC), and system on chip (SOC), which offer the highest amount of logic density and fast processing.

The logic gates IC is under SSI, combinational logic circuits are part of MSI, and microprocessor is the part of LSI and VLSI level.

Advantages of VLSI technology:

- Size of the device will be reduced
- Weight of the device will be reduced
- Power consumption is reduced
- The development time will be less.
- The cost will be reduced for end users.
- Reliability of the entire system will be more

1.2 Digital System Design

Digital system design is the process of designing or developing systems that represent information using a binary system. It's easier to store, reproduce, transmit, and manipulate digital data and cheaper/easier to design such systems.

The system which can process a continuous range of values and signals is known as the analog system. In an analog device, data is represented by physical variables, whereas in a digital device, the data is represented by physical variables. In daily life, we experience so many analog systems such as analog watches, voltmeter, speedometer, CRO readings, etc. Analog systems have been used for so many decades and have various applications in the electronics industry. The digital system deals with a discrete range of values and signals. For example, a digital weighing machine indicates a

Table 1.2 Comparison between analog and digital system

Parameters	Analog System	Digital System
Nature of signal	Continuous signals	Discrete signals
Accuracy and precision	Less accurate	More accurate
Noise effect	Much affected	Less effected
Memory and storage	Memory is not available	It has memory
Fabrication of IC	Analog IC is difficult to fabricate	Digital IC is easy to fabricate
Size	Large in size	Small and compact
Reliable	Reliability is less	More reliable
Examples	Voltmeter, power supply, CRO	Counters, digital meter, computers

weight of 47.6 kg, as compared to the analog weighing machine. Reading an analog device involves human error and approximation.

1.3 Boolean Algebra

Boolean algebra was first introduced by George Boole in 1847. In contrast with elementary algebra, the value of input variables in Boolean algebra is truth values, i.e., true or false which are represented in the digital system as 0 and 1. Boolean algebra is a mathematical tool to analyze digital circuits.

The main motive of logic design is to minimize and simplify the logic so that a minimum number of gates or wires are used, which will cause–effect the cost, power dissipation, and speed of that logic circuit. Using Boolean algebraic rules, the equations can be simplified and logic complexity can be further reduced.

1.3.1 Boolean Variables

In Boolean algebra, the variables may have value either 0 or 1. The number of variables may-be infinite. For example, if $Y = A + B$, there are 3 variables involved but the value of A and B will be either 0 or 1. Although Boolean algebra is not only restricted to the base 2 system. The interpretation of binary variables used for Boolean algebra is

Table 1.3 Value of Boolean variables

Value	Interpretation
0	LOW, FALSE, NO, OFF
1	HIGH, TRUE, YES, ON

1.3.2 Boolean Operators

Three operators are used in Boolean algebra: a compliment, addition, and multiplication. These are the building blocks of Boolean algebra.

Table 1.4 Boolean operators

Boolean Operators	Interpretation	Examples
NOT	Compliment, inversion	$\overline{Y} = A'$
AND	Union, logical addition	$Y = AB$
OR	Intersection, logical multiplication	$Y = A + B$

1.3.3 Boolean Laws and Theorems

In this section, basic Boolean laws are provided, using Boolean laws, simplified Boolean expressions can be carried out and used to simplify and minimize the logic circuits. Furthermore, Boolean laws and theorems are proved.

Table 1.5 Boolean laws and theorems

S. Nos.	Boolean Law		Description
	Law	**Dual Property**	
1	$\bar{\bar{A}} = A$		Involution Law
	OR law	**AND law**	
2	$0 + A = A$	$1 \cdot A = A$	Addition and multiplication of identity element
3	$1 + A = 1$	$0 \cdot A = 0$	Dominance
4	$A + A = A$	$A \cdot A = A$	Idempotent
5	$A + \bar{A} = 1$	$A \cdot \bar{A} = 0$	Compliment
	Commutative Law		
6	$A + B = A + B$	$A \cdot B = B \cdot A$	The order does not change the result
	Associative Law		
7	$A + (B + C) = (A + B) + C$	$A \cdot (B \cdot C) = (A \cdot B) \cdot C$	Associative law
	Distributive Law		
8	$A(B + C) = AB + AC$	$A + BC = (A + B) \cdot (A + C)$	Dual is not valid in normal algebra

(*Continued*)

Table 1.5 Continued

S. Nos.	Boolean Law		Description
	Law	**Dual Property**	
		Theorem	
9	$A + AB = A$	$A(A + B) = A$	Absorption
10	$A + \bar{A}B = A + B$	$A(\bar{A} + B) = AB$	Degenerate-reflect or redundant law
		Census Theorem	
11	$AB + \bar{A}C + BC = AB + \bar{A}C$	$(A + B) \cdot (\bar{A} + C) \cdot (B + C) = (A + B) \cdot (\bar{A} + C)$	
		Demorgan Theorem	
12	$\overline{A + B} = \bar{A} \cdot \bar{B}$	$\overline{AB} = \bar{A} + \bar{B}$	Demorgan

1.4 Combinational and Sequential Design

In the combinational logic circuits, the output depends on the present input. But there are many areas where previous output effects the present output, i.e., concept of memory arises to store the previous output. So, the combinational circuits when used along with memory are known as sequential circuits. The past/previous output is provided to the input with the help of feedback.

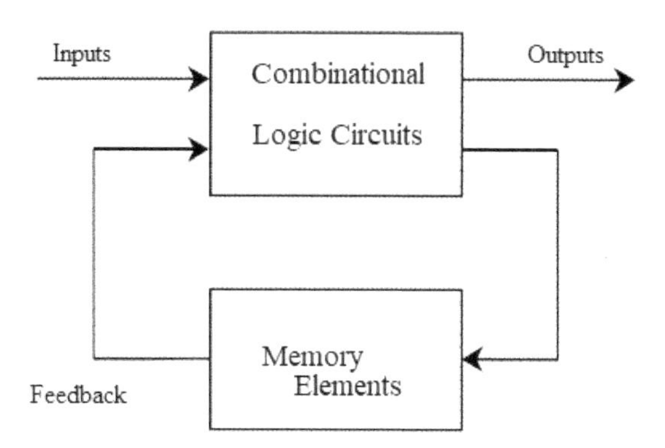

Figure 1.1 Block diagram of sequential circuits.

The block diagram of sequential logic circuits is depicted in Figure 1.1, where it has been shown that outputs depend on the inputs and previous output, which are stored in memory elements. These memory elements are connected with primary inputs through feedback channels.

1.4.1 Types of Sequential Logic Circuits

A sequential circuit is known by a time sequence of all the input and output parameters. Its behavior is specified by the value of signals at the discrete instance of time. Depending on the timing of circuits, the sequential circuits are further classified into three types, as shown in Figure 1.2.

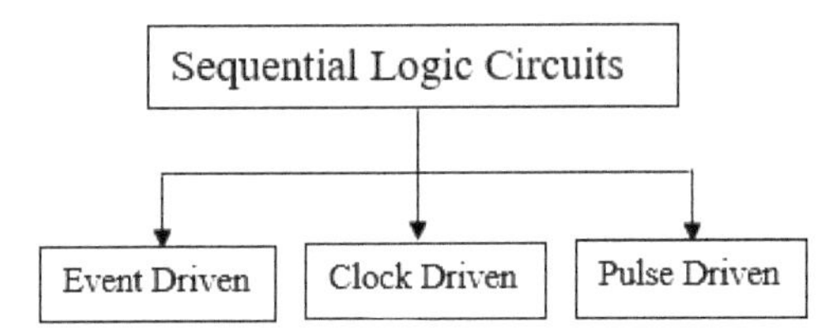

Figure 1.2 Types of sequential logic circuits.

The types of sequential logic circuits are:

1. Event driven, also known as asynchronous circuits
2. The clock is driven, also known as synchronous circuits
3. Pulse is driven

In the event-driven or asynchronous circuits, the output of logic circuit changes as soon as there is drift/change in the input. So, it is known as event-driven or asynchronous circuits, as shown in Figure 1.3.

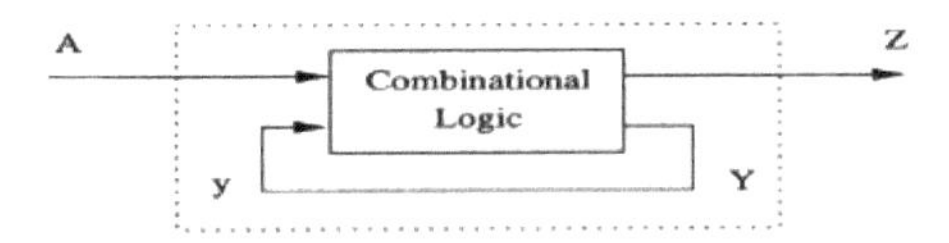

Figure 1.3 Asynchronous sequential circuits.

In the clock-driven or synchronous circuits, a signal named as the clock signal is used to determine or control the exact time, when output can change. It means that the clock is an added feature, which controls the operation of the logic circuit. It is shown in Figure 1.4.

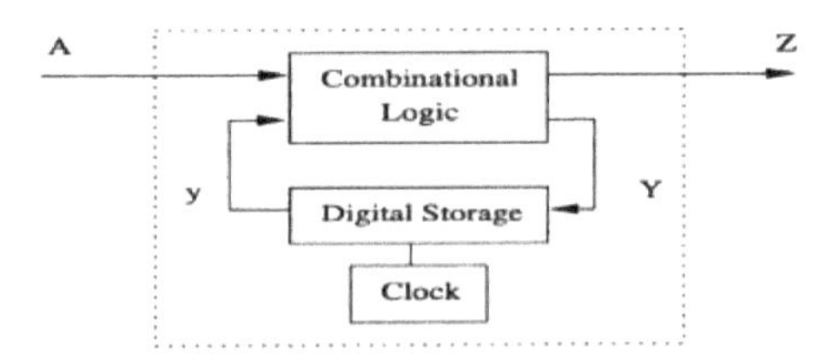

Figure 1.4 Synchronous sequential circuits.

The pulse driven sequential logic circuits is the combination of two, where output depends on the triggering of pulses.

1.5 Digital-Logic Family

The digital electronics has a variety of logic gates and circuit elements which realize various logic functions. A group of electronic devices or elements that possess a similar range of characteristics and built by the same manufacturing technique is known as a logic family. Logic gates and memory devices are fabricated as ICs. With the advancement in technology, the level of integration enhances. The various active or passive components are interconnected within the chip to form a digital circuit. The chip is mounted on the metal or plastic package, the connections are welded externally to form an IC.

1.5.1 Classification of Logic Families

Based on manufacturing technology, ICs are broadly categorized as

(a) Bipolar families
(b) Unipolar families

Based on internal characteristics and fabrication process, they are grouped in different logic families, as follows:

(a) RTL: Resistor transistor logic
(b) DTL: Diode transistor logic
(c) TTL: Transistor–transistor logic
(d) ECL: Emitter-coupled logic
(e) I^2L: Integrated injection logic
(f) MOS: Metal–oxide–semiconductor

Figure 1.5 shows the classification of digital logic families. Out of these logic families, RTL and DTL are obsolete now, and TTL and ECL are mostly used in design and realization. For low power consumption devices such as calculators, CMOS is a good option.

1.5.2 Nomenclature of the Logic Family

The manufacturer of ICs follow a standard numbering scheme, in which suffix, middle, and prefix parts have some significance. For example, the 7400 series IC and its derivatives are shown in Figure 1.6.

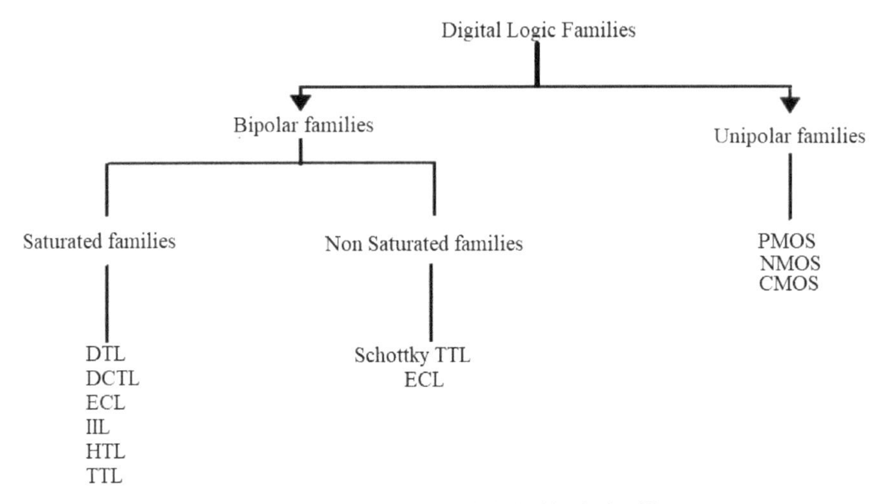

Figure 1.5 Classification of digital logic families.

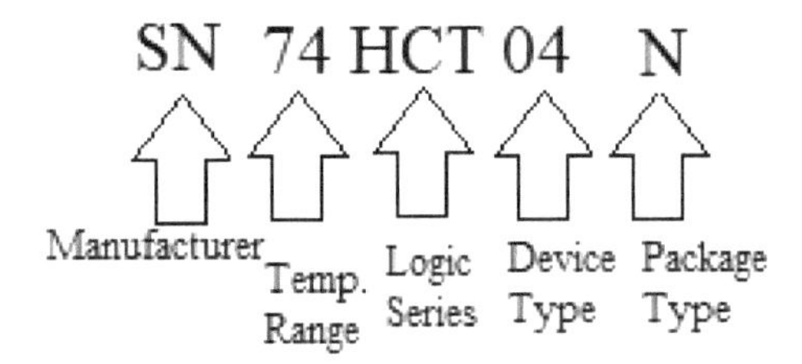

Figure 1.6 IC number SN74HCT04N.

Significance of IC number:

(a) Manufacturer: This code consists of two alphabets, e.g., SN is used for Texas instruments, HEF for Mullard/Philips, DM for National semiconductors, and S for Signetics.
(b) Temperature range: This code consists of two numeric. 74 signifies temperature range 0–70°C commercial, 54 military –55–125°C.
(c) Logic series: This signifies subfamily. 7400 series are most widely used in consumer applications.
(d) Device type: This indicates a device type of function. For example, 04 indicates hex/inverter.
(e) Package type: It signifies the package type of IC, for example, N is for plastic dual in-line, W is for the ceramic flat-pack, D surface mounted plastic package.

Technical Questions with Solutions

- **Ques: 1. How will you define bit and byte in a digital system?**

Bit: It is a fundamental unit of the binary number system. It can be represented as 0 or 1.

Byte: The group of 8 bits is called a byte. It estimates the size of computer memory.

- **Ques: 2. What are the limitations of the digital circuit?**

The real world is analog-based. So, DAC (digital-to-analog converters) is used which operates on digital information and convert digital blocks to real-world analog form.

- **Ques: 3. The presence and absence of students in a class represent 1 or 0, respectively. Represent the binary form of the attendance sheet.**

P	P	A	A	P
A	P	P	A	A
A	A	A	P	P

Row 1: 11001; Row 2: 01100; Row 3: 00011

- **Ques: 4. Express generally a number in any digital number system.**

If a number is having n digits and d values per position, r represents the radix of a number, and X is the number, then it can be generally displayed as

$$X = d_n r^n + d_{n-1} r^n - 1 + \cdots + d_1 r^1 + d_0 r^0$$

- **Ques: 5. Explain the terms: (a) MSB and (b) LSB.**

The bit which is having the highest power of radix is known as the most significant bit (MSB). It is at the extreme left.

The bit which is having the lowest power of radix is known as the least significant bit (LSB). It is at the extreme right.

- **Ques: 6. Define the significance of the radix point.**

The radix point is the point which separates the integer and fraction parts when expressed in the form of positional radix. For example, in a binary system, the number can be expressed as

$$2^4\, 2^3\, 2^2\, 2^1\, 2^0 \cdot 2^{-1}\, 2^{-2}\, 2^{-3}\, 2^{-4}$$

$$\uparrow$$

Radix point

- **Ques: 7. Give the daily life example of an analog system and a digital system.**

Analog system: analog voltmeter, analog ammeter, analog clock, analog weighing scale, old radio, old telephone handset, record player, etc.

Digital system: digital clock, digital multimeter, home security system, digital alarm clock, compact disc player, digital wristwatch, digital scale, counter, etc.

- **Ques: 8. What is the positional weight system?**

In the positional weight system, the position of a digit determines the weight. It is associated with a radix point. Any number system can be represented using a positional weight system.

- **Ques: 9. In the numbers 4150 and 1450, determine the weight of 4.**

In 4150, the weight of 4 is 1000 and in 1450 the weight of 4 is 100.

- **Ques: 10. Explain the true complement of a binary number.**

The true complement of the binary system is known as 2's complement. In this, subtract each digit of number from $r - 1$ and then add 1 to LSB.

- **Ques: 11. Which logic gate satisfies the operation of 1's complement?**

The NOT gate having an inverted operation satisfies the 1's complement function.

- **Ques: 12. Explain the two-part conversion method.**

While converting decimal to a binary number, the two-part method is used. Where integer part of the decimal number is divided by 2 again and again. Mention 0 if the remainder is not present, otherwise write 1. The first division represents LSB and the last division represents MSB.

- **Ques: 13. Explain 10 as a signed magnitude number.**

The decimal number $10 = (1010)_2$
Positive number $+10 = (0\ 1010)_2$
Negative number $-10 = (1\ 1010)_2$

- **Ques: 14. Why hexadecimal numbers are preferred mostly?**

Representing the number in the binary system takes so many bits. It produces a long string. It is difficult to read and understand such long numbers. The hexadecimal numbers are formed by grouping 4 bits. The data becomes concise and simple. So, the hexadecimal number system is preferred over other number systems.

- **Ques: 15. What is the minimum to maximum range for two-digit, three-digit, and four-digit hexadecimal numbers?**

The two-digit hexadecimal number is $(FF)_{16}$. The range of numbers is from 0 to (r^2 – 1), i.e., 0–to 255. The three-digit hexadecimal number is $(FFF)_{16}$. The range of numbers is from 0 to (r^2 – 1), i.e., 0 to 4095. The four-digit hexadecimal number is $(FFFF)_{16}$. The range of numbers is from 0 to (r^2 – 1), i.e., 0–65535.

- **Ques: 16. Find X, Y, and radix of following equation:**

$$(XY)r = (25)_{10} \quad \text{and} \quad (YX)r = (31)_{10} \quad \text{and} \quad Y - X = 1$$

Since the base or radix is r, so

$$Xr^1 + Yr^0 = 25$$

or $$Xr + Y = 25 \quad (1)$$

and $$Yr^1 + Xr^0 = 31$$

or $$Yr + X = 31 \quad (2)$$

Also, $$Y = X + 1 \quad (3)$$

From Equations (1–3)

$$X = 3, \quad Y = 4, \quad \text{and} \quad r = 7$$

- **Ques: 17. Explain the different formats of binary floating-point numbers.**

The different formats for floating-point numbers are

(i) Single-precision: it has 32 bits
(ii) Double precision: it has 64 bits
(iii) Extended precision: it has 80 bits

- **Ques: 18. What is the advantage of the octal number system?**

For expressing a long sequence of binary numbers, the octal number system is an attractive and concise one where a grouping of 3 binary bits is created.

- **Ques: 19. Define nibble and word.**

Nibble: The group of 4 bits is called a nibble. It is half the size of a byte.

Word: The combination of two or more bits formulates a word. Word length can be 16, 32, or 64 bits, etc.

- **Ques: 20. What is a two-stage operation in digital systems?**

In digital systems, two-state operation means level low and level high. The voltage having value 5 V is represented by logic 1 and it is the part of level high. Whereas 0 V represents logic 0 or level low.

- **Ques: 21. A seven-bit Hamming code as received is 1111101. Check if it is correct. If not find the correct code if even parity is used.**

Solution:

D7	D6	D5	P4	D3	P2	P1
1	1	1	1	1	0	1

The bits 4, 5, 6, 7 have even number of 1s. Hence, no error.
The bits 2, 3, 6, 7 have an odd number of 1s. Hence, error.
The bits 1, 3, 5, 7 have even number of 1s. Hence, no error.
The error is in bit 2 position.
The correct code is 1111111.

- **Ques: 22. Elaborate the operation of NAND gate as negative OR gate.**

The operation of the NAND gate and OR gate is shown in the following table:

The truth table of OR and NAND gate

INPUT		OUTPUT	
A	B	OR	NAND
0	0	0	1
0	1	1	1
1	0	1	1
1	1	1	0

From the above table, we can easily deduce that the operation of the NAND gate and OR gate are in a complementary manner or we can say that, the aspect of the NAND gate operation is negative OR.

Further, it can be shown with the help of logic gate

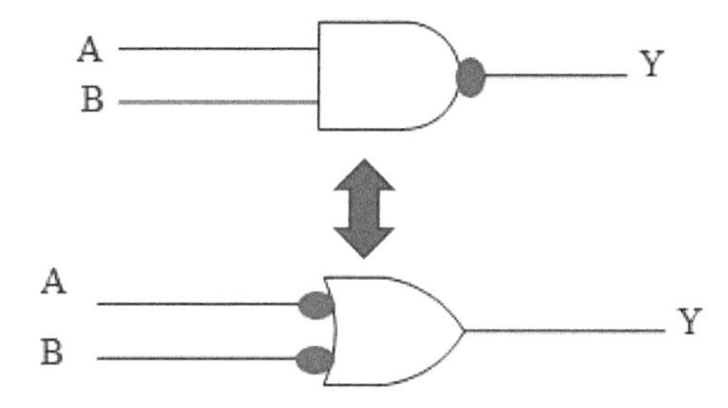

As, we can see bubbles are present at input terminals, so input variables become $\bar{A}$ and $\bar{B}$. The OR of these two variables are

$\bar{A} + \bar{B} = \overline{AB}$, which is a NAND gate.

- **Ques: 23. Prove that NAND is a universal logic gate**

A universal logic gate is that property of a logic gate, which enables it to create basic gate AND, OR and NOT from it.

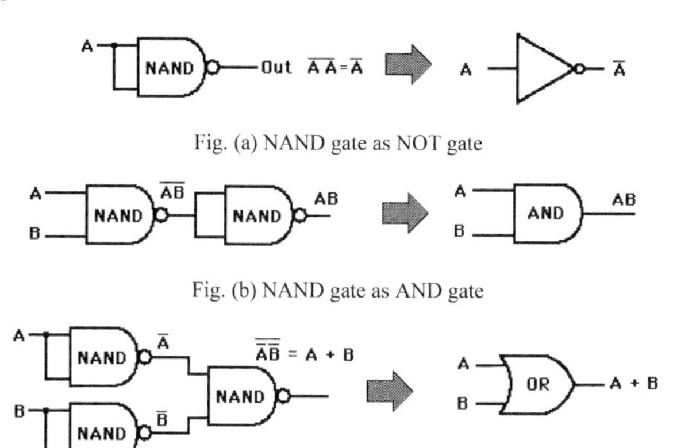

Fig. (a) NAND gate as NOT gate

Fig. (b) NAND gate as AND gate

Fig. (c) NAND gate as OR gate

Hence proved, NAND is a universal logic gate, because using only NAND gate other basic gates are derived successfully.

- **Ques: 24. Prove that NOR is a universal logic gate.**

A universal logic gate is that property of a logic gate, which enables it to create basic gate AND, OR, and NOT from it.

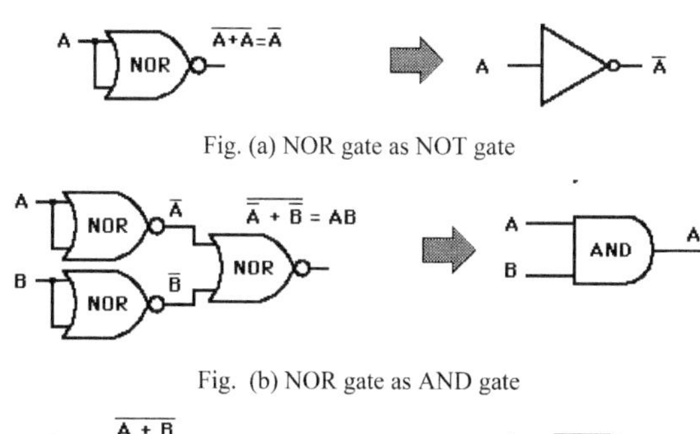

Fig. (a) NOR gate as NOT gate

Fig. (b) NOR gate as AND gate

Fig. (c) NOR gate as OR gate

Hence proved, NOR is a universal logic gate, because using only NOR gate other basic gates are derived successfully.

- **Ques: 25. Prove that A + 1 = 1**

Solution:

$$\begin{aligned} A + 1 &= 1.(A + 1) \\ &= (A + \bar{A}).(A + 1) \quad \text{(1 can be written as } \bar{A} + A) \\ &= (A \cdot A) + (A \cdot 1) + (\bar{A} \cdot A) + (\bar{A} \cdot 1) \\ &= A + A + 0 + \bar{A} \quad (A \cdot A = A) \\ &= \overline{A + A} \\ &= 1 \end{aligned}$$

- **Ques: 26. Prove that complement of 0 is 1.**

Solution:

$$\begin{aligned} \bar{0} &= \bar{0} + 0 \\ \bar{0} &= 1(\text{Because } A + \bar{A} = 1) \end{aligned}$$

Hence proved.

- **Ques: 27. Prove the Idempotency law: A + A = A**

Solution:

$$\begin{aligned} \text{LHS} &= (\text{A} + \text{A}) \cdot 1 \\ &= (A + A) \cdot (A + \bar{A}) \\ &= A \cdot 1 \\ &= A \end{aligned}$$

Hence proved.

- **Ques: 28. Prove the Annulment law:** $\bar{\bar{A}} = A$

Solution:
Using Boolean laws $A + \bar{A} = 1,\ A\bar{A} = 0,\ \bar{A}\bar{\bar{A}} = 0,\ \bar{A} + \bar{\bar{A}} = 1$, we can prove this law

$$\begin{aligned} \text{LHS}\bar{\bar{A}} &= \bar{\bar{A}} + 0 \\ &= \bar{\bar{A}} + A \cdot \bar{A} \\ &= (\bar{\bar{A}} + A) \cdot (\bar{\bar{A}} + \bar{A}) \\ &= (\bar{\bar{A}} + A) \cdot 1 \end{aligned}$$

$$
\begin{aligned}
&= (\bar{\bar{A}} + A) \cdot (A + \bar{A}) \\
&= (A + \bar{\bar{A}}) \cdot A + \bar{A}) \\
&= A + (\bar{\bar{A}} \cdot \bar{A}) \quad \text{(distributive property)} \\
&= A
\end{aligned}
$$

Hence proved.

- **Ques: 29. Prove the Absorption law:** $A + AB = A$

Solution:
Using Boolean law $1 + A = 1$, we can prove absorption law

$$
\begin{aligned}
\text{LHS} A + AB &= A(1 + B) \\
&= A \cdot 1 \\
&= A
\end{aligned}
$$

Hence proved.

- **Ques: 30. Prove degenerate-reflection law.**
  - $A + \bar{A}B = A + B$

Solution:
Using Boolean law $A + \bar{A} = 1$, we can prove degenerate-reflection law

$$
\begin{aligned}
\text{LHS } A + \bar{A}B &= (A + \bar{A}) \cdot (A + B) \\
&= 1 \cdot (A + B) \\
&= A + B
\end{aligned}
$$

Hence proved.

- **Ques: 31. Prove the distributive property:** $A(B + C) = AB + AC$

$$A + BC = (A + B) \cdot (A + C)$$

Solution:
Using Identity and Idempotent Boolean laws,

$$
\begin{aligned}
\text{RHS } (A + B) \cdot (A + C) &= A \cdot A + A \cdot C + B \cdot A + B \cdot C \\
&= A + AC + AB + BC \\
&= A(1 + C) + AB + BC \\
&= A + AB + BC
\end{aligned}
$$

$$= A(1 + B) + BC$$
$$= A + BC$$

Hence proved.

- **Ques: 32. Using Perfection Induction law prove commutative property.**

$$A \cdot B = B \cdot A$$

Solution:
Using perfect induction law, values of AB are taken as 00, 01, 10, and 11

A	B	A · B	B · A
0	0	0	0
0	1	0	0
1	0	0	0
1	1	1	1

Hence proved.

- **Ques: 33. Using Perfection Induction law prove associative property**

$$A \cdot (B \cdot C) = (A \cdot B) \cdot C$$

Solution:
Using perfect induction law, values of ABC are taken as 000, 001, 010, 011, 100, 101, 110, and 111 and perform the desired function.

A	B	C	A · (B · C)	(A · B) · C
0	0	0	0	0
0	0	1	0	0
0	1	0	0	0
0	1	1	0	0
1	0	0	0	0
1	0	1	0	0
1	1	0	0	0
1	1	1	1	1

Hence proved.

- **Ques: 34. Prove Consensus Theorem:** $AB + \bar{A}C + BC = AB + \bar{A}C$

Solution:

Using distributive, commutative, identity, and complement law

$$\begin{aligned}\text{LHS } AB + \bar{A}C + BC &= AB + \bar{A}C + 1 \cdot BC \\ &= AB + \bar{A}C + (A + \bar{A}) \cdot BC \\ &= AB + \bar{A}C + ABC + \bar{A}BC \\ &= AB + ABC + \bar{A}C + \bar{A}CB \\ &= AB \cdot 1 + ABC + \bar{A}C \cdot 1 + \bar{A}CB \\ &= AB(1 + C) + \bar{A}C(1 + B) \\ &= AB \cdot 1 + \bar{A}C \cdot 1 \\ &= AB + \bar{A}C\end{aligned}$$

Hence proved.

- **Ques: 35. Verify Demorgan's theorem:**
  (a) $\overline{A + B} = \bar{A} \cdot \bar{B}$
  (b) $\overline{AB} = \bar{A} + \bar{B}$

Solution:

(a) Using perfect induction law, we can verify Demorgan's theorem, here the values of AB are taken as 00, 01, 10, and 11.

A	B	$A + B$	$\overline{A + B}$	$\bar{A}$	$\bar{B}$	$\bar{A} \cdot \bar{B}$
0	0	0	1	1	1	1
0	1	1	0	1	0	0
1	0	1	0	0	1	0
1	1	1	0	0	0	0

Hence proved $\overline{A + B} = \bar{A} \cdot \bar{B}$

(b) Using perfect induction law, we can verify Demorgan's theorem, here the values of AB are taken as 00, 01, 10, and 11

A	B	AB	$\overline{AB}$	$\bar{A}$	$\bar{B}$	$\bar{A} + \bar{B}$
0	0	0	1	1	1	1
0	1	0	1	1	0	1
1	0	0	1	0	1	1
1	1	1	0	0	0	0

Hence proved $\overline{AB} = \bar{A} + \bar{B}$

- **Ques: 36. What will be the output Y of a given gate, if input B is low?**

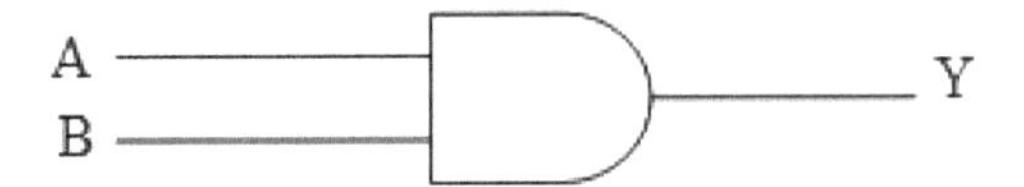

Solution:
If the input B is low, it means the value of B is "0." So, Y will also be low because it is AND gate and output of AND gate is high only in the case when all the inputs are high. But here one input is low, so Y is also low. Moreover, if B is low, i.e., 0, $Y = A \cdot B = A \cdot 0 = 0$, so the value of Y is 0.

- **Ques: 37. Give the IC numbers of all the logic gates.**

Solution:
All the logic gates in the digital system have integrated circuits. The IC numbers are

Logic Gate	IC Number
AND	7408
OR	7432
NOT	7404
NAND	7400
NOR	7402
EX-OR	7486
EX-NOR	74266

- **Ques: 38. Illustrate the pulsed operation of an inverter or NOT gate.**

Solution:
NOT gate or inverter performs an inversion operation.

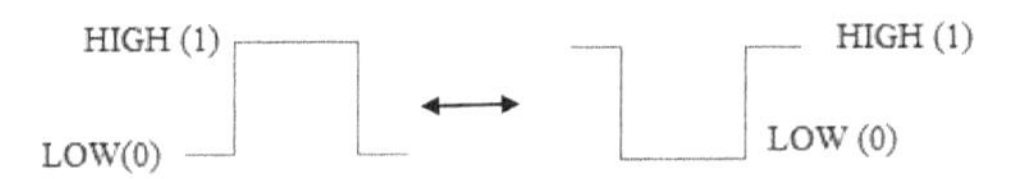

- **Ques: 39. What is even and an odd number of negations?**

Solution:

The even number of negations means no negation.
The odd number of negations means a single negation.

- **Ques: 40. Draw the pin diagram of AND, OR, and NOT gate.**

Solution:

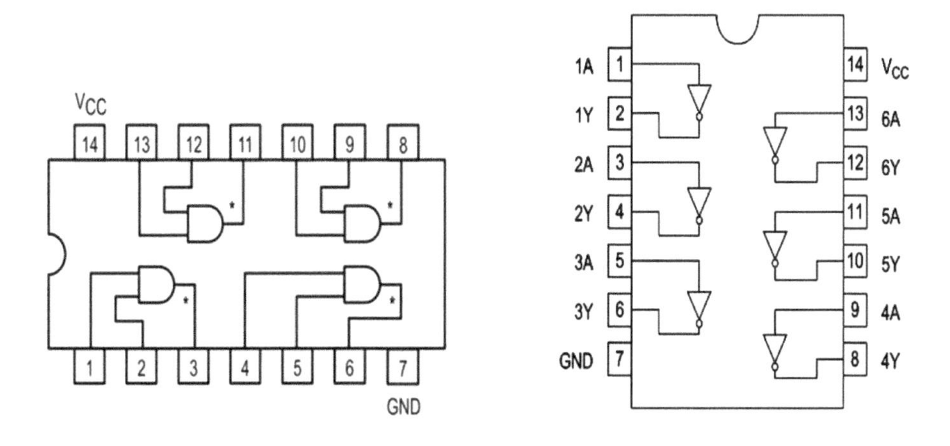

(a) AND gate (b) NOT gate

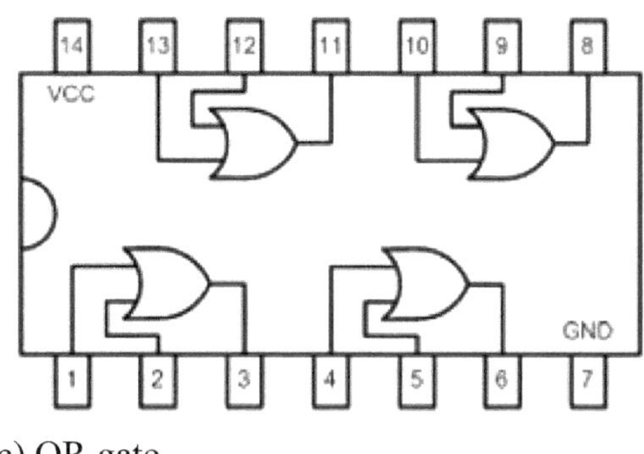

(c) OR gate

- **Ques: 41. Draw the pin diagram of the NAND and NOR gate.**

Solution:

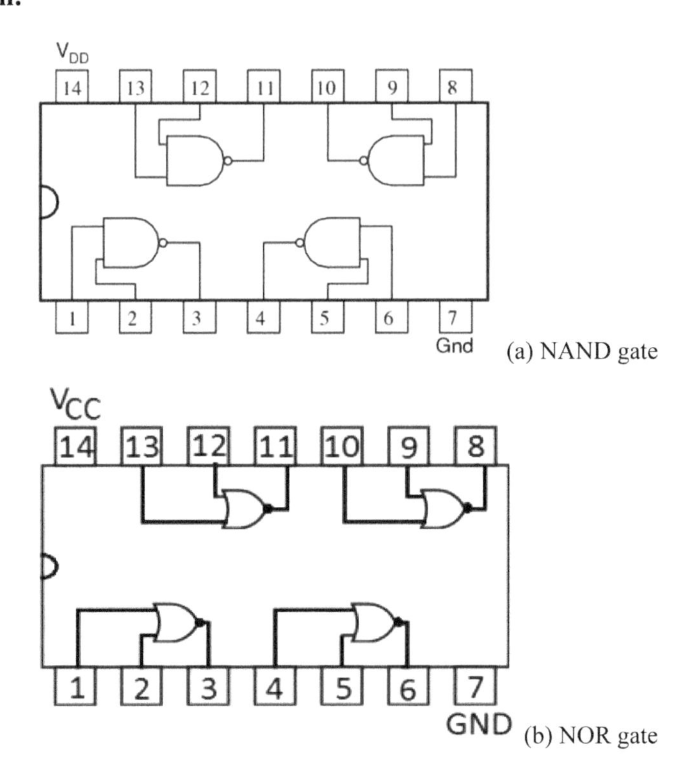

(a) NAND gate

(b) NOR gate

- **Ques: 42. Prove that $A + \bar{A}B = A + B$.**

Solution:

$$\begin{aligned}
A + \bar{A}B &= (A + AB) + \bar{A}B \qquad \text{Because } (A + AB = A) \\
&= A \cdot A + A \cdot B + \bar{A}B \qquad (A \cdot A = A) \\
&= A \cdot A + A \cdot B + 0 + \bar{A}B \qquad (A \cdot \bar{A} = 0) \\
&= A \cdot A + A \cdot B + A \cdot \bar{A} + \bar{A} \cdot B \\
&= (A + \bar{A}) \cdot (A + B) = 1 \cdot (A + B) = A + B
\end{aligned}$$

Hence proved.

- **Ques: 43. Prove that** $(A + B)(A + C) = A + BC$.

Solution:

$$\begin{aligned}(A+B)(A+C) &= AA + AC + AB + BC \qquad \text{(distributive law)}\\ &= A + AC + AB + BC \qquad (A \cdot A = A)\\ &= A(1 + C) + AB + BC\\ &= A + AB + BC \qquad (1 + C = 1)\\ &= A(1 + B) + BC\\ &= A + BC \qquad (1 + B = 1)\end{aligned}$$

Hence proved.

- **Ques: 44. Prove that** $A(\bar{A} + C)(\bar{A}B + \bar{C}) = 0$

Solution:

$$\begin{aligned}A(\bar{A}+C)(\bar{A}B+\bar{C}) &= (A\bar{A} + AC)(\bar{A}B + \bar{C})\\ &= (0 + AC)(\bar{A}B + \bar{C}) \qquad (A \cdot \bar{A} = 0)\\ &= AC\bar{A}B + AC\bar{C}\\ &= 0 \cdot BC + A \cdot 0\\ &= 0\end{aligned}$$

- **Ques: 45. Give the law of dualization in Boolean algebra.**

Solution:
The law of dualization is also known as De Morgan's theorem. It formulates the relationship between N(AND) and N(OR) functions that allows one type of function to be implemented using a different type of gates.
(a) $\overline{A + B} = \bar{A} \cdot \bar{B}$
(b) $\overline{A \cdot B} = \bar{A} + \bar{B}$

- **Ques: 46. Reduce the expression** $A + B(\overline{C + \overline{DE}})$.

Solution:
By applying De Morgan's law:

$$\begin{aligned}A + B(\overline{C + \overline{DE}}) &= A + B(\overline{C + \bar{D} + \bar{E}})\\ &= A + B(\bar{C}DE)\\ &= A + B\bar{C}DE\end{aligned}$$

- **Ques: 47. Prove that** $\overline{\overline{AB} + \bar{A} + AB} = 0.$

Solution:

$$\begin{aligned}
\overline{\overline{AB} + \bar{A} + AB} &= \bar{A} + \bar{B} + \bar{\bar{A}} + AB \text{ (De Morgan's law)} \\
&= \bar{A} + \bar{\bar{B}} + AB \\
&= \bar{A} + \bar{\bar{\bar{B}}} + A \quad (A + \bar{A}B = A + B) \\
&= \overline{1 + \bar{\bar{B}}} \qquad (A + \bar{A} = 1) \\
&= \bar{1} = 0 \qquad (1 + \bar{B} = 1)
\end{aligned}$$

- **Ques: 48. Simplify the Boolean expression:** $[A\bar{B}(C + BD) + \bar{A}\bar{B}]C.$

Solution:

$$\begin{aligned}
[A\bar{B}(C + BD) + \bar{A}\bar{B}]C &= [(A\bar{B}C + A\bar{B}BD) + \bar{A}\bar{B}]C \\
&= (A\bar{B}C + 0 + \bar{A}\bar{B})C \qquad (\bar{B}B = 0) \\
&= (A\bar{B}CC + \bar{A}\bar{B}C) \\
&= (A\bar{B}C + \bar{A}\bar{B}C) \qquad (C \cdot C = C) \\
&= \bar{B}C(A + \bar{A}) \qquad A + \bar{A} = 1) \\
&= \bar{B}C
\end{aligned}$$

- **Ques: 49.** Convert the following expression to both AND and OR gate.

$$Y = \overline{A \cdot B} + \bar{C} \cdot D$$

Solution:

(i) First, we will convert $\overline{A \cdot B}, \bar{C} \cdot D$ into AND expression using De Morgan's theorem

$$\overline{A \cdot B} + \bar{C} \cdot D = \overline{A \cdot B \cdot \quad \cdot \bar{C}D}$$

(ii) Now, we will convert $\overline{A \cdot B}, \bar{C} \cdot D$ into OR expression using De Morgan's theorem

$$\begin{aligned}
\overline{A \cdot B} &= \bar{A} + \bar{B} \\
\bar{C}D &= C + \bar{D}
\end{aligned}$$

Hence, $\overline{A \cdot B} + \bar{C} \cdot D = \bar{A} + \bar{B} + C + \bar{D}$

- **Ques: 50. Simplify using De Morgan's law**

$$F = \bar{A} + \bar{B} + \overline{\overline{(A \cdot \bar{C}} + B)} + C$$

Solution:
Using De Morgan's law, the given Boolean function can be simplified.
$\bar{A}+\bar{B}+\overline{\overline{(A \cdot \bar{C}} + B)}+C$ can be minimized by simplifying the inside Boolean relations

$$\overline{A \cdot \bar{C} + \bar{B}} = A \cdot \bar{C}.B$$

$$\bar{A} + \bar{B} + C = \overline{A \cdot B. \cdot \bar{C}}$$

The total expression is $\overline{A \cdot B. \cdot \bar{C}} + A \cdot \bar{C}.B$

If $A \cdot \bar{C}.B$ is X, then $\overline{A \cdot B. \cdot \bar{C}}$ is $\bar{X}$

As, $X + \bar{X} = 1$

Hence, $F = 1$

- **Ques: 51. Apply De Morgan's law on** $F = \overline{(A + B) + C}$

Solution:
By applying De Morgan's law:

$$\overline{(A + B) + C} = \overline{(A + B)}.\bar{C}$$
$$= \bar{A}.\bar{B}.\bar{C}$$

- **Ques: 52. Apply De Morgan's law on** $F = \overline{(A + B)CD + E + F}$

Solution:
By applying De Morgan's law:

$$\overline{(A + B)CD + E + F} = [\overline{(A + B)CD}]\overline{(E + F)}$$
$$= (\overline{AB} + \bar{C} + \bar{D})\bar{E}\bar{F}$$

- **Ques: 53. Simplify the expression:**

$$F = B\bar{C}\bar{D} + \bar{A}BD + ABD + BC\bar{D} + \bar{B}CD + \bar{A}\bar{B}\bar{C}D + A\bar{B}\bar{C}D$$

Solution:

$$B\bar{C}\bar{D} + \bar{A}BD + ABD + BC\bar{D} + \bar{B}CD + \bar{A}\bar{B}\bar{C}D + A\bar{B}\bar{C}D$$
$$= BD(A + \bar{A}) + B\bar{D}(C + \bar{C}) + \bar{B}CD + \bar{B}\bar{C}D(A + \bar{A})$$
$$= BD + B\bar{D} + \bar{B}CD + \bar{B}\bar{C}D$$

$$= B(D + \bar{D}) + \bar{B}D(C + \bar{C})$$
$$= B + \bar{B}D \qquad (D + \bar{D} = 1)$$
$$= B + D \qquad (A + \bar{A}B = A + B)$$

- **Ques: 54. Simplify the expression:**

$$F = \overline{\overline{X + Y} + \bar{Z}}$$

Solution:
By applying the Boolean laws and theorems:

$$F = \overline{\overline{X + Y} + \bar{Z}}$$
$$= \overline{\overline{X + Y}}.\bar{\bar{Z}}$$
$$= (X + Y).Z$$

- **Ques: 55. Simplify the expression:**

$$F = \overline{\overline{AB} + \bar{A} + AB}$$

Solution:
By applying the Boolean laws and theorems:

$$\overline{\overline{AB} + \bar{A} + AB} = \bar{A} + \bar{B} \bar{+} \bar{A} + AB \qquad (\bar{A} + AB = \bar{A} + B)$$
$$= \bar{A} + \bar{B} \bar{+} \bar{A} + B$$

By arranging the terms

$$= \overline{\bar{A} + \bar{A} + \bar{B} + B}$$
$$= \bar{1} = 0$$

- **Ques: 56. Simplify** $F = AB + A(B + C) + B(B + C)$

Solution:

$$AB + A(B + C) + B(B + C)$$
$$= AB + AB + AC + BB + BC \qquad (B. \cdot B = B)$$
$$= AB + AC + B + BC$$
$$= AB + AC + B(1 + C) \qquad (1 + C = 1)$$
$$= AB + AC + B$$
$$= AB + B + AC$$

$$= B(A+1) + AC$$
$$= B + AC$$

So, $F = B + AC$

- **Ques: 57. What do you mean by minterm and maxterm?**

Solution:
The product term in SOP (sum of products) is minterm and the sum term in POS (product of sums) is maxterm.

- **Ques: 58. Convert the given function into standard SOP form**

$$F = AB + A\bar{C} + BC$$

Solution:

$$\begin{aligned} F &= AB + A\bar{C} + BC \\ &= AB(C+\bar{C}) + A(B+\bar{B})\bar{C} + (A+\bar{A})BC \\ &= ABC + AB\bar{C} + A\bar{B}\bar{C} + ABC + \bar{A}BC \\ &= AB\bar{C} + A\bar{B}\bar{C} + \bar{A}BC + ABC \end{aligned}$$

- **Ques: 59. Convert the given function into standard POS form**

$$F = (A+B)(B+C)(A+C)$$

Solution:

$$\begin{aligned} F &= (A+B)(B+C)(A+C) \\ &= (A+B+C\bar{C})(A\bar{A}+B+C)(A+B\bar{B}+C) \\ &= (A+B+C)(A+B+\bar{C})(A+B+C)(\bar{A}+B+C) \\ &\quad (A+B+C)(A+\bar{B}+C) \\ &= (A+B+C)(A+B+\bar{C})(A+\bar{B}+C)(\bar{A}+B+\bar{C}) \end{aligned}$$

- **Ques: 60. Simplify the Boolean function:** $F = A\bar{B}C + \bar{A}\bar{B}C$

Solution:

$$\begin{aligned} F &= A\bar{B}C + \bar{A}\bar{B}C \\ &= \bar{B}C(A+\bar{A}) \\ &= \bar{B}C \cdot 1 \\ &= \bar{B}C \end{aligned}$$

- **Ques: 61. Explain the concept of two forms of a Boolean expression**

Solution:
Boolean expressions can be written as (a) SOP form: Sum of products form (b) POS form: Products of sum form

- **Ques: 62. Prove that SOP and POS are complementary.**

Solution:
Let us take an expression $AB + BC$ in the SOP form.

$$\begin{aligned}\overline{AB + BC} &= \overline{AB}.\overline{BC} \\ &= (\bar{A} + \bar{B}).(\bar{B} + \bar{C})\end{aligned}$$

So, $\overline{SOP} = POS$

- **Ques: 63. Explain the significance and need for Boolean algebra.**

Solution:
Using binary numbers 0 and 1, Boolean algebra is helpful to minimize the logic functions and solve the Boolean expressions. In such a way, while implementing the Boolean expression, a lesser number of gates will be used which consumes less power and the circuit will be fast. So, Boolean algebra is useful in optimizing the logic circuits.

- **Ques: 64. What is the method to represent minterms and maxterms?**

Solution:
The minterms are product terms of SOP and maxterms are sum terms of POS. Minterms are represented by "1" and maxterms are represented by "0." The Boolean expression must convert into standard canonical form before representing in terms of minterms and maxterms.

In SOP function $F = ABC + AB\bar{C}$, the minterms are $ABC,\ AB\bar{C}$

In POS function $F = (A + B + C).(A + \bar{B} + C)$, the max terms are $(A + B + C), (A + \bar{B} + C)$

- **Ques: 65. If you have two exclusive OR gates, how can it possible that one will act as an inverter and others will act as a buffer?**

Solution:
If we fix one of the inputs of the exclusive OR gate to 1, it will act as an inverter. In another case, if we fix one of the inputs of an exclusive OR gate to 0, it will act as a buffer.

- **Ques: 66. If you have a two-input NAND gate, tell the way how it will act as an inverter.**

Solution:
The possible solution is short both inputs of NAND gate and apply a single input to it, it will act as an inverter.

- **Ques: 67. State and prove De Morgan's theorem.**

Solution:
According to De Morgan's law, the complement of the sum of two numbers $(\overline{A+B})$ is equal to the product of both numbers in their complement form, i.e., $\bar{A}.\bar{B}$. Similarly, the complement of the product of two numbers A and B, i.e., $\overline{AB}$ is equal to the sum of both numbers in their complement form, i.e., $\bar{A}+\bar{B}$.
(a) $\overline{A+B}=\bar{A}.\bar{B}$
(b) $\overline{AB}=\bar{A}+\ \bar{B}$

(a) Using perfect induction law, we can verify De Morgan's theorem; here, the values of AB are taken as 00, 01, 10, and 11.

A	B	$A+B$	$\overline{A+B}$	$\bar{A}$	$\bar{B}$	$\bar{A}.\bar{B}$
0	0	0	1	1	1	1
0	1	1	0	1	0	0
1	0	1	0	0	1	0
1	1	1	0	0	0	0

Hence proved $\overline{A+B}=\bar{A}.\bar{B}$
(b) Using perfect induction law, we can verify De Morgan's theorem, here the values of AB are taken as 00, 01, 10, and 11.

A	B	AB	$\overline{AB}$	$\bar{A}$	$\bar{B}$	$\bar{A}+\bar{B}$
0	0	0	1	1	1	1
0	1	0	1	1	0	1
1	0	0	1	0	1	1
1	1	1	0	0	0	0

Hence proved $\overline{AB}=\bar{A}+\ \bar{B}$

- **Ques: 68. Find out the minterms if** $F = AB + ABC + AB\bar{C}$.

Solution:
The first step is the conversion of a given function into a standard SOP form.

$$\begin{aligned}F &= AB(C+\bar{C}) + ABC + AB\bar{C}\\ &= ABC + AB\bar{C} + ABC + AB\bar{C}\\ &= ABC + AB\bar{C}\\ &= 111, 110\\ &= \text{m7, m6 are the desired minterms}\end{aligned}$$

- **Ques: 69. Simplify using K-map** $F = \sum m(0, 2, 4, 8, 10, 12) + d(5)$

Solution:
As it is the case of minterms, so enter "1" where terms are present in the given expression.

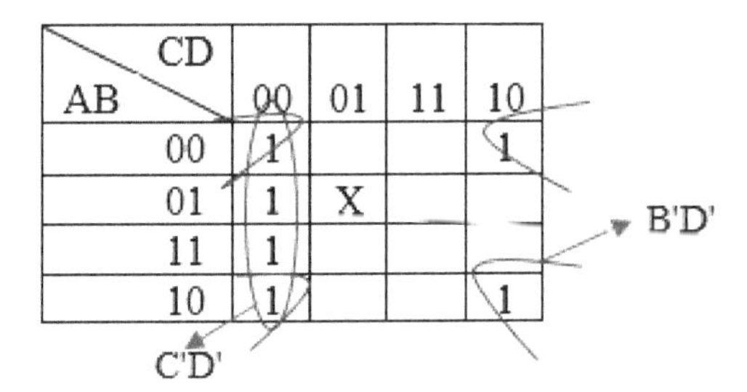

So, $F = \bar{C}\bar{D} + \bar{B}\bar{D}$

- **Ques: 70. Simplify using K-map** $F = \prod M(0, 1, 3, 5, 6, 7, 10, 14, 15) + d(8)$

Solution:
As it is the case of maxterms, so, enter "0" where the term is present in the expression

AB \ CD	00	01	11	10
00	0	0	0	
01		0	0	0
11			0	0
10				0

A+B+C
A+D'
B'+C'
A'+C'+D

So, $F = (A + B + C).(A + \bar{D}).(\bar{B} + \bar{C}).(\bar{A} + \bar{C} + D)$

- **Ques: 71. Simplify using K-map** $F = \sum m(0, 4, 5, 6, 10) + d(1, 2, 14)$.

Solution:
As it is the case of minterms, so enter "1" where terms are present in the given expression.

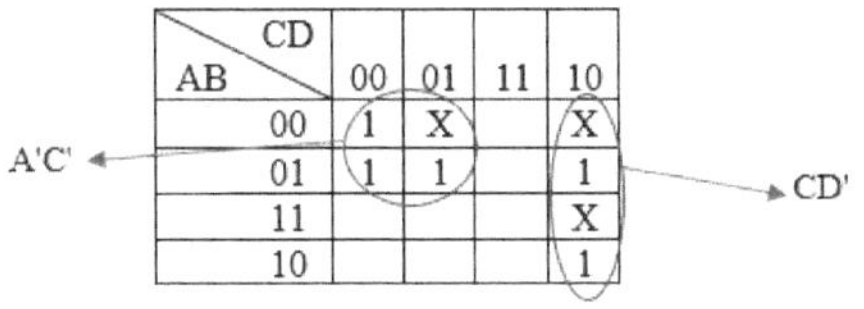

AB \ CD	00	01	11	10
00	1	X		X
01	1	1		1
11				X
10				1

A'C'
CD'

So, $F = \bar{A}\bar{C} + C\bar{D}$

- **Ques: 72. Simplify using K-map** $F = \prod M(0, 1, 6, 8) + d(9, 11, 13, 15)$.

Solution:
As it is the case of maxterms, so, enter "0" where the term is present in the expression

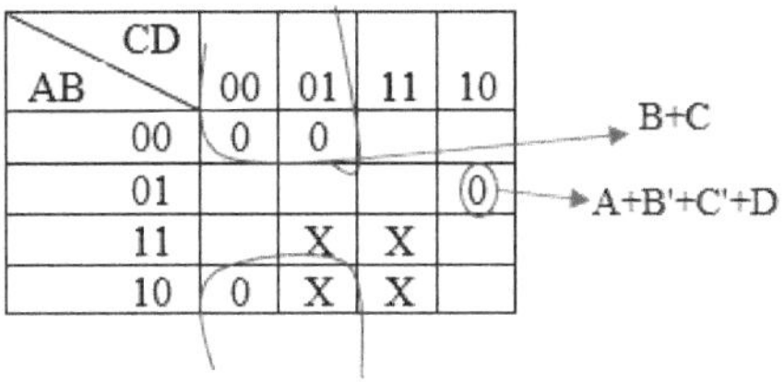

AB \ CD	00	01	11	10
00	0	0		
01				0
11		X	X	
10	0	X	X	

B+C
A+B'+C'+D

Here, note that we cannot make a pair of only "X." Don't care are only to extend the pair or create the pair.

$$\text{So, } F = (B + C).(A + \bar{B} + C + \bar{D})$$

- **Ques: 73. What do you understand by the tabulation method?**

Solution:
The tabulation method is used for exhaustive searching purposes. It specifies and creates the minimum expression to solve Boolean expression for a greater number of variables.

- **Ques: 74. What is the term "Don't care variable?" Specify its need.**

Solution:
Don't care variables are the part of K-map simplification. While creating SOP or POS, minterms and max terms are used, "1" and "0" need to be filled in

K-map against the position of respective minterm or maxterm. They don't care variables are often represented as "X" in K-map respective cells. Their basic use is to extend the existing group or pair. The designer can consider X as 0 or 1, as output for each combination.

- **Ques: 75. Give the logical expression for exclusive OR and exclusive NOR gate.**

Solution:

$$A\ ex - or\ B = A\bar{B} + \bar{A}B$$
$$A\ ex - nor\ B = AB + \bar{A}\bar{B}$$

- **Ques: 76. What do you understand by minimized expression?**

Solution:
An expression that is simplified using Boolean algebra and other simplification techniques, i.e., K-map, is known as minimized expression. It has the minimum number of logic gates. If the number of gates is less, the wires and interconnections will be minimum, which will further reduce cost and power consumption.

- **Ques: 77. What are the basic constraints for a chip designer?**

Solution:
While designing a circuit, the designer needs to take care of four parameters:

1. Size
2. Cost
3. Speed
4. Power consumption

As the integration level is upgraded to ULSI (ultra-large-scale integration), where millions of transistors are installed at a small chip. It becomes a major issue for the designer to produce fast, cheap, and tiny-sized chips that consume less power.

- **Ques: 78. Give the application of exclusive OR gate**

Solution:
Exclusive OR gate is used in a 1-bit magnitude comparator. It gives output = 1 when both inputs are equal in magnitude.

- **Ques: 79. Design a 4–16 decoder using a 3–8 decoder.**

Solution:
Using two 3–8 decoders, a 4–16 decoder can be designed in the following manner.

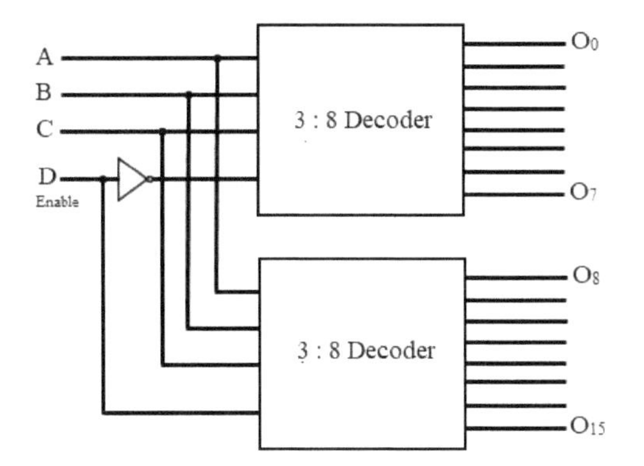

4–16 decoder using 3–8 decoder.

- **Ques: 80. Represent full adder using half-adder.**

The full adder can be represented as the combination of two half-adders as a figure.

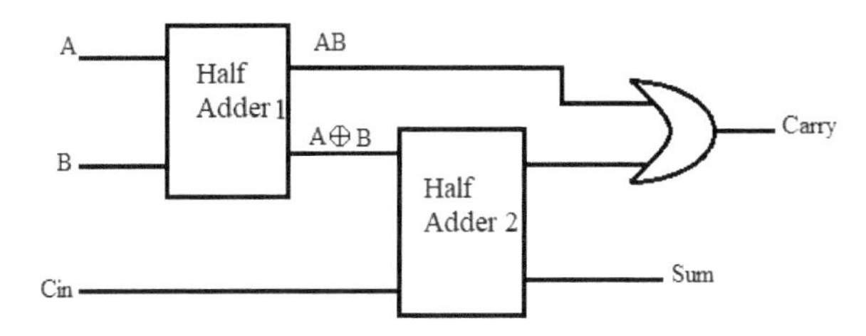

Full adder using half-adder.

- **Ques: 81. Design full subtractor using half-subtractor.**

The full subtractor can be designed using half-subtractor as a figure.

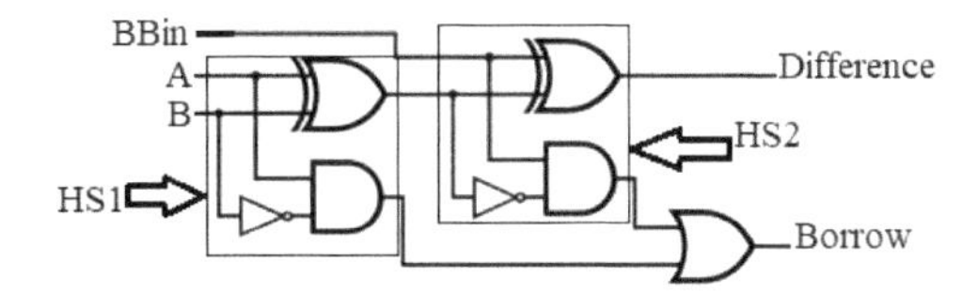

Full subtractor using half-subtractor.

- **Ques: 82. Implement full adder using 3:8 decoder.**

Solution:

The full adder is used for an additional purpose, where three bits are added to produce the sum and carry as output.

Table. The truth table of full adder

Inputs			Outputs	
A	B	C	Sum	Carry
0	0	0	0	0
0	0	1	1	0
0	1	0	1	0
0	1	1	0	1
1	0	0	1	0
1	0	1	0	1
1	1	0	0	1
1	1	1	1	1

Using the truth table, we can write

$$Sum = \sum m(1, 2, 4, 7)$$
$$Carry = \sum m(3, 5, 6, 7)$$

As the number of inputs is 3, and outputs are 8. So, 3:8 decoder is used to realize the full adder.

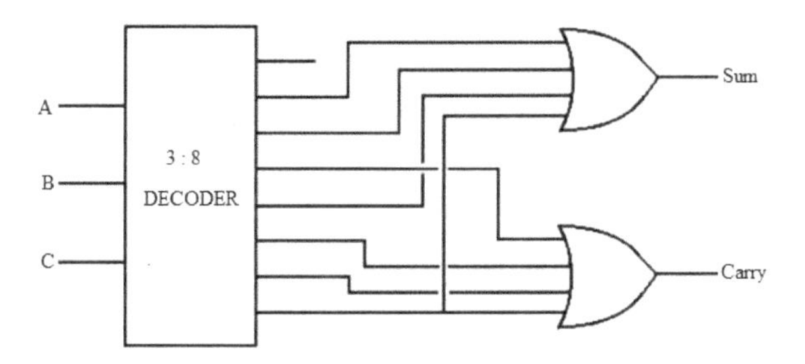

Full adder using 3:8 decoder.

- **Ques: 83. How can the function of the combinational logic circuit be specified?**

Solution:

The three main ways of specifying the function of a combinational logic circuit are:

- Boolean algebra: This forms the algebraic expression showing the operation of the logic circuit for each input variable either true or false that results in a logic “1” output.

- Truth table: A truth table defines the function of a logic gate by providing a concise list that shows all the output states in tabular form for each possible combination of input variables that the gate could encounter.
- Logic diagram: This is a graphical representation of a logic circuit that shows the wiring and connections of each logic gate, represented by a specific graphical symbol, that implements the logic circuit.
- **Ques: 84. Why half-adders are called half-adder?**

Solution:

The half-adder adds two bits and generates carry. But it does not take care of the input. That means if we use half-adder in a circuit and if there is a carry generated in the previous state then it won't consider it.

That's why we have to use full adder as it adds two bits along with consideration of carrying in from the previous state. Hence, it is named full adder. Now, a full adder circuit can be obtained by connecting two half-adders. This makes its name the half-adder. Two half-adders make a full adder.

- **Ques: 85. Full adder using two half-adders.**

Solution:

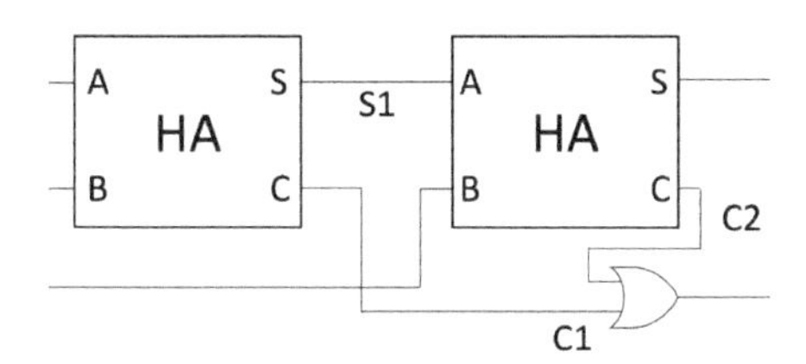

Full adder using half-adders.

- **Ques: 86. What is the application of multiplexer?**

Solution:

The multiplexers are used as one method of reducing the number of logic gates required in a circuit design or when a single data line or data bus is required to carry two or more different digital signals. For example, a single 8-channel multiplexer.

- **Ques: 87. Implement the following Boolean function using 8:1 MUX.**

$$F(A, B, C, D) = \sum m(0, 1, 4, 6, 7, 8, 11, 13, 15)$$

Solution:

The 8:1 Mux will have three select lines, so B, C, and D are fixed as select lines.

	I0	I1	I2	I3	I4	I5	I6	I7
A'	0	1	2	3	4	5	6	7
A	8	9	10	11	12	13	14	15
	1	A'	0	A	A'	A	A'	1

The same can be realized using a logic diagram:

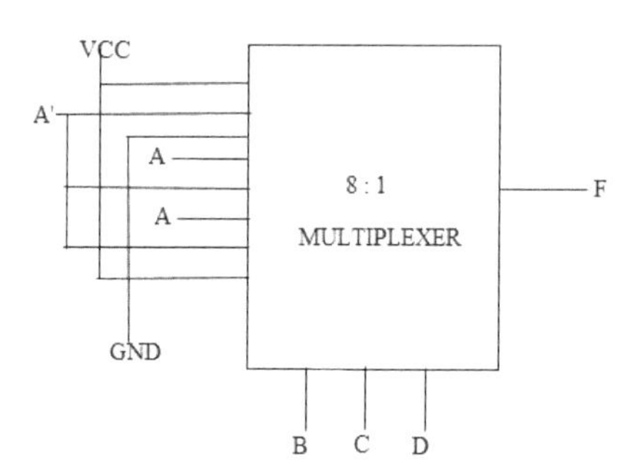

Implementation using 8:1 Mux.

- **Ques: 88. Design 5:32 decoder using 2:4 decoder and 3:8 decoder.**

Solution:

The 5:32 decoder can be designed using 3:8 decoder and 2:4 decoder, by the following way:

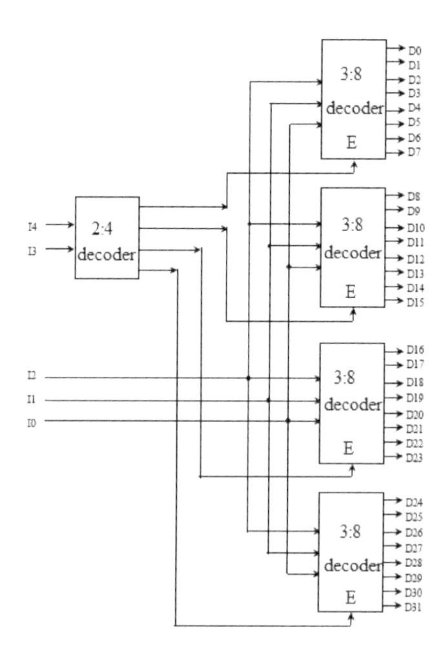

Decoder using 2:4 and 3:8 decoder.

- **Ques: 89. Give the practical examples of multiplexer and demultiplexer.**

Solution:
Mobile communication, selection of movie screens, home stereo system, etc.

- **Ques: 90. Give the practical life examples of encoder and decoder.**

Solution:
Encoders are used in calculators, transmitters, linear motion to digital signal converters.
Decoders are used in receivers, calculators, etc.

- **Ques: 91. Give the IC numbers for decoders.**

Solution:

IC Numbers	Functions
74138	Octal Decoder
7442	BCD Decoder
74154	Hex Decoder
7447	BCD to seven segment decoder

- **Ques: 92. Draw the pin diagram for octal decoder 74138.**

Solution:
The pin diagram of 74138 is

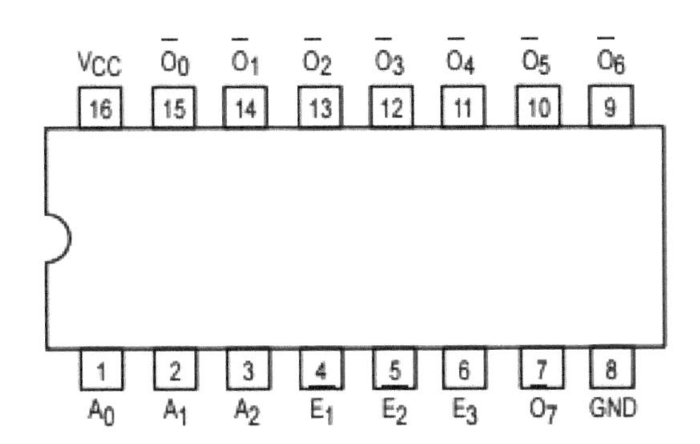

Pin diagram of an octal decoder.
where E is enabled pin and O is an output pin.

- **Ques: 93. What is the function of the data selector?**

Solution:
The data selector is also known as a digital multiplexer, which selects one output from the range of many inputs, with the help of control signals or select lines.

- **Ques: 94. List any three available multiplexers ICs.**

Solution:
The multiplexers that are available in the form of ICs are

Functions	Devices
Quad 2-input	74157
Dual 8-input	74153
8-input	74151
16-input	74150

- **Ques: 95. Differentiate between gray code and binary code.**

Solution:
The gray code varies by only one bit from one entry to next and from last entry to first. The basic operation that is performed in conversion from binary to gray code is exclusive OR gate operation.

- **Ques: 96. What is the significance of code converters?**

Solution:
The code converters convert the coded number into another form that is more usable by a computer or a digital system. For example, binary to gray code converter, BCD to seven segment code converter, binary to excess-3 code converter, etc.

- **Ques: 97. What is logic diagram representation?**

Solution:
The logic diagram helps visualize the gate implementation of the Boolean expressions or equations.

- **Ques: 98. Explain the binary to BCD conversion.**

Solution:
Using adder circuits, binary numbers can be converted into BCD numbers. The conversion process is as follows:

1. Represent the value of each BCD number in terms of the binary bit.
2. Add all the binary representation of bits that are 1s in BCD number.

3. The result represents the binary equivalent of the BCD number in which the binary equivalent of each BCD number represents the weight of that bit, within the total BCD number.

- **Ques: 99. Differentiate between combinational and sequential circuits.**

Solution:

S. Nos.	Combinational Circuit	Sequential Circuit
1	A circuit in which output depends only on the present inputs.	A circuit in which output depends on present inputs as well as past inputs.
2	There is no memory element or feedback system.	Memory elements and feedback systems are present.
3	Easy to design	Sequential circuits are comparatively harder to design.
4	It requires more hardware for designing purposes.	Less hardware is required.
5	Combinational circuits are expensive circuits.	Sequential circuits contain less hardware, so cheaper in comparison to combinational circuits.
6	Combinational circuits are faster in speed.	Due to memory or delay elements, sequential circuits are slower
7	The behavior is defined by the set of output functions only.	The behavior is defined by the set of output and the next state functions only.
8	Designer has less flexibility. For example: parallel adder	Sequential circuits are more flexible due to memory elements. For example: counters

- **Ques: 100. Differentiate between synchronous sequential circuits and asynchronous sequential** circuits.

Solution:

S. Nos.	Synchronous sequential circuits	Asynchronous sequential circuits
1	In synchronous sequential circuits, the behavior is defined from the knowledge of its signals at the discrete instant of time, i.e., input signals can affect the memory elements upon activation of the clock pulse.	In asynchronous sequential circuits, the behavior depends upon the order in which the input changes and state of circuits, i.e., change in the input signal can affect memory element at any instant of time.
2	Memory elements are clocked flip-flops.	Memory elements are either upclocked flip-flops or delay elements.
3	Synchronous sequential circuits are synchronized by clock signals.	Asynchronous sequential circuits are not synchronized by clock signals.
4	Slower in comparison to asynchronous sequential circuits	Asynchronous sequential circuits are faster because they do not wait for the clock signals.
5	Design of synchronous sequential circuits is easier.	Asynchronous sequential circuits are difficult to design.
6	The maximum operating speed depends on the time delay and synchronization.	The speed of operation is independent of clock or time delay.

- **Ques: 101. What is edge triggering and level triggering of flip-flop?**

Solution:
Edge triggered flip-flops are enabled by the positive or negative edge of digital waveform and level triggered flip-flops are enabled by a logic high or low level. The transition of waveform or clock pulse from HIGH to LOW is a negative edge and from LOW to HIGH is a positive edge waveform.

- **Ques: 102. What is synchronous and asynchronous input of flip-flop?**

Solution:

Synchronous input of a flip-flop that does not affect the flip-flop's outputs unless a clock pulse is applied. Flip-flop's inputs and clock are synchronous inputs.

The input of a flip-flop that changes the outputs of the flip-flops immediately, without waiting for a pulse at the clock input (preset and clear). The preset is an asynchronous set function. Clear is an asynchronous reset function.

- **Ques: 103. What is setup time and hold time in flip-flops?**

Solution:

Setup time (t_s) is the time required for synchronous inputs of a flip-flop to be stable when a clock pulse is applied.

Hold time (t_h) is the time that the synchronous inputs of a flip-flop must remain stable after the active clock transition is finished.

- **Ques: 104. What is the pulse width of a flip-flop?**

Solution:

Pulse width (t_w) is the minimum time required for an active level pulse applied to the clock, as measured from the center point of the leading edge of the pulse to the center point of the trailing edge.

- **Ques: 105. What are rise time and fall time of flip-flops?**

Solution:

The time required to change the voltage level from 10% to 90% is known as rising time (t_r) and the time required to change the voltage level from 90% to 10% is known as fall time (t_f).

- **Ques: 106. Show the clock parameters of flip-flop graphically.**

Solution:

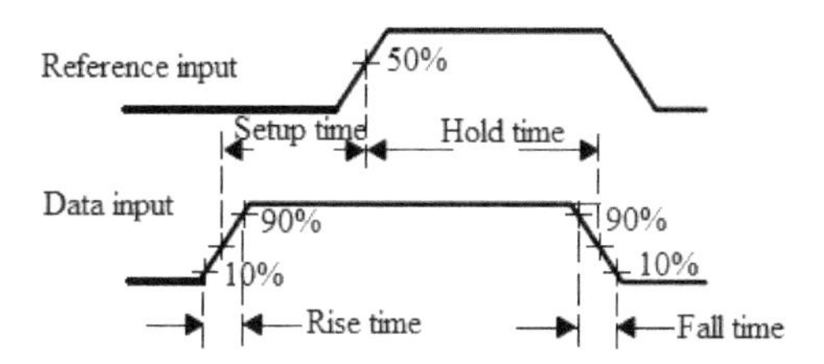

Flip-flop clock parameters.

- **Ques: 107. What is the difference between an asynchronous counter and a synchronous counter?**

Solution:

S.Nos.	Asynchronous counter	Synchronous counter
1	The flip-flops are arranged in such a manner that output of first flip-flop becomes the clock of next flip-flop.	In the synchronous counter, there is no connection between the output of flip-flop and the clock of the flip-flop.
2	All the flip-flops are not clocked simultaneously.	Al the flip-flops are run by a single clock pulse.
3	Even if the number of states is more, the design is simple.	The design will be more complex with the increase in the number of states.
4	The speed of the circuit is slow because waiting or timing delay will be more.	The speed of synchronous counters is fast because they are operated by a common clock pulse.

- **Ques: 108. What is the universal shift register?**

Solution:

A universal shift register is often bidirectional as well. It can operate with any combination of serial and parallel inputs and outputs (i.e., serial in serial out, serial in parallel out, parallel in serial out, and parallel in parallel out). So, the shift registers which have both shift and parallel load capabilities are referred to as a universal shift register.

- **Ques: 109. What are the applications of shift registers?**

Solution:

1. A SISO register is used to introduce a time delay in digital signals.
2. A SISO register is used as serial to parallel converter.
3. A PISO register is used as a parallel to serial converter.
4. Shift registers are also used as shift register counters, for example, ring counter and Johnson counter.

- **Ques: 110. What is a flip-flop?**

Solution:
Flip-flops are a basic sequential logic device that has two stable states. It can serve as a memory device. It is also called as a bistable multivibrator.

- **Ques: 111. Give some applications of clocked SR flip-flop.**

Solution:

1. Clocked SR flip-flops are used in calculators and computers.
2. SR flip-flops are widely used in electronic products.

- **Ques: 112. Give some applications of D flip-flop.**

Solution:

1. It is used as a temporary memory device.
2. D flip-flops are wired together to form shift registers and storage registers.
3. D flip-flops are used in digital counters.

- **Ques: 113. What is toggling in JK flip-flop?**

Solution:
In JK flip-flop, when both J and K are 1, the repeated clock pulse causes the output to turn off and turn on. This repeatedly off-on action is like a toggle switch and hence it is called toggling.

- **Ques: 114. Give some applications of flip-flops.**

Solution:

1. Parallel data storage
2. Data transfer
3. Frequency division
4. Counting

- **Ques: 115. What is a ripple counter?**

Solution:
In the ripple counter, the transition state of LSB flip-flop triggers the clock of the next flip-flop and so on. Time delay results from the rippling of the count from LSB to MSB.

- **Ques: 116. What is the modulus of a counter?**

Solution:
The modulus of a counter is the number of different states, the counter must go through, to complete the counting cycle.

- **Ques: 117. Compare synchronous counters and ripple counters.**

Solution:

S.Nos.	Synchronous counter	Ripple counter
1	A common clock triggers all flip-flops simultaneously.	An external clock pulse triggers the first flip-flop. The second flip-flop is triggered by the output of the first flip-flop.
2	Synchronous counters are triggered by either a positive or negative clock edge.	The polarity of the clock is essential to trigger the flip-flops.

- **Ques: 118. What are the properties of shift registers?**

Solution:

1. Acts as temporary memory.
2. It has shifting characteristics, i.e., shifts the numbers to the left on the display each time when a new digit is pressed on the keyboard.

- **Ques: 119. What are the applications of shift registers?**

Solution:

1. Serial to parallel converter
2. Parallel to serial converter
3. Time delay element
4. Counters

- **Ques: 120. How the shift registers are used for time delay?**

Solution:
The SISO register is used to introduce a time delay in the digital signals. The time delay t_d is related to frequency as

$$Td = N.\frac{1}{fc}$$

where N is the number of states and f_c is the clock frequency.

- **Ques: 121. How ring counter can be converted to Johnson counter?**

Solution:
When the complemented output of the last flip-flop in ring counter is connected to the first flip-flop, it will work as Johnson counter.

- **Ques: 122. If data is 1101, then how many clock pulses are needed for the SISO register to perform its operation?**

Solution:
The data is 4 bit long. In SISO the operation is performed serially. So, for data input 4 clock pulses are needed and for data output 4 clock pulses are needed. A total of 8 clock pulses are needed by SISO to perform its operation.

- **Ques: 123. If data is 101, then how many clock pulses are needed for the SIPO register to perform its operation?**

Solution:
The data is 3 bit long. In SIPO the read operation (data input) is performed serially and write operation (data output) is performed parallelly. So, for data input 3 clock pulses are needed and for data output 1 clock pulse is needed. A total of 4 clock pulses are needed by SIPO to perform its operation.

- **Ques: 124. If data is 01, then how many clock pulses are needed for the PISO register to perform its operation?**

Solution:
The data is 2 bit long. In PISO the read operation (data input) is parallel and write operation is performed serially. So, for data input 1 clock pulse is needed and for data output 2 clock pulse is needed. A total of 3 clock pulses is needed by PISO to perform its operation.

- **Ques: 125. If data is 111, then how many clock pulses are needed for the PIPO register to perform its operation?**

Solution:
The data is 3 bit long. In PIPO the operation is performed parallelly. So, for data input 1 clock pulse is needed and for data output 1 clock pulse is needed. A total of 2 clock pulses is needed by PIPO to perform its operation.

- **Ques: 126. What is the bidirectional shift register?**

Solution:
A register that is capable of shifting in both directions, i.e., left to right and right to left, is known as a bidirectional shift register.

- **Ques: 127. What is the decade counter? How many flip-flops are required in the design of the decade counter?**

Solution:

A decade counter is also known as MOD-10 counter, it counts from 0 to 9 and then resets its operation. The number of flip-flops required is 4.

- **Ques: 128. Design Mod 4 synchronous counter using JK flip-flop and implement it.**

Solution:

Step 1: Mod4 indicates the number of states, i.e., 0, 1, 2, and 3.
Step 2: The required number of flip-flops is 2. Because of $4 = 2^2$.
Step 3: Excitation table of JK flip-flop

Q(t)	Q(t+1)	J	K
0	0	0	X
0	1	1	X
1	0	X	1
1	1	X	0

Step 4: State table

Present inputs		Next state		Flip-flop inputs			
Q_A	Q_B	Q_{A+1}	Q_{B+1}	J_A	K_A	J_B	K_B
0	0	0	1	0	X	1	X
0	1	1	0	1	X	X	1
1	0	1	1	X	0	1	X
1	1	0	0	X	1	X	1

Step 5: Logic expression using K-map

QA \ QB	0	1
0		1
1	X	X

$JA = QB$

QA \ QB	0	1
0	X	X
1		1

$KA = QB$

QA \ QB	0	1
0	1	X
1	1	X

$JB = 1$

QA \ QB	0	1
0	X	1
1	X	1

$KB = 1$

Step 6: Counter design using JK flip-flop

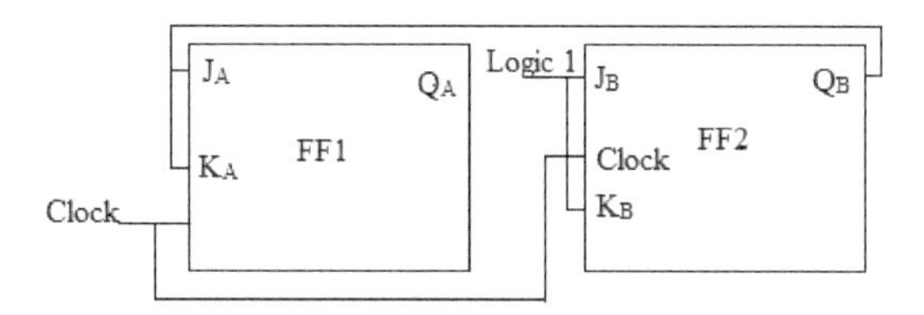

Mod4 synchronous counter using JK flip-flop.

- **Ques: 129. Design 3-bit up/down counter using T flip-flop.**

Solution:

An up/down counter is also known as a bidirectional counter. As both operations up and down have been performed in a single circuit, so its operation is controlled by $up/\overline{down}$. When this signal is low, decremented/down operation will be performed and when this signal is at a high level, up/incremented operation will be performed.

State table:

Input	Present state			Next state			Flip-flop inputs		
$up/\overline{down}$	Q_C	Q_B	Q_A	Q_{C+1}	Q_{B+1}	Q_{A+1}	T_C	T_B	T_A
0	0	0	0	1	1	1	1	1	1
0	0	0	1	0	0	0	0	0	1
0	0	1	0	0	0	1	0	1	1
0	0	1	1	0	1	0	0	0	1
0	1	0	0	0	1	1	1	1	1
0	1	0	1	1	0	0	0	0	1
0	1	1	0	1	0	1	0	1	1
0	1	1	1	1	1	0	0	0	1
1	0	0	0	0	0	1	0	1	1
1	0	0	1	0	1	0	0	0	1
1	0	1	0	0	1	1	0	1	1
1	0	1	1	1	0	0	1	0	1
1	1	0	0	1	0	1	0	1	1
1	1	0	1	1	1	0	0	0	1
1	1	1	0	1	1	1	0	1	1
1	1	1	1	0	0	0	1	0	1

The logical expressions can be found using K-map

U_DQ_C \ Q_BQ_A	00	01	11	10
00	1			
01	1			
11			1	
10			1	

$$TC = \overline{UD}.\overline{QB}.\overline{QA} + UD.QB.QA$$

U_DQ_C \ Q_BQ_A	00	01	11	10
00	1			1
01	1			1
11		1	1	
10		1	1	

$$TB = \overline{UD}.\overline{QA} + UD.QA$$

U_DQ_C \ Q_BQ_A	00	01	11	10
00	1	1	1	1
01	1	1	1	1
11	1	1	1	1
10	1	1	1	1

$$TC = 1$$

After extracting the logical expression, the 3-bit up/down synchronous counter is designed using T flip-flop.

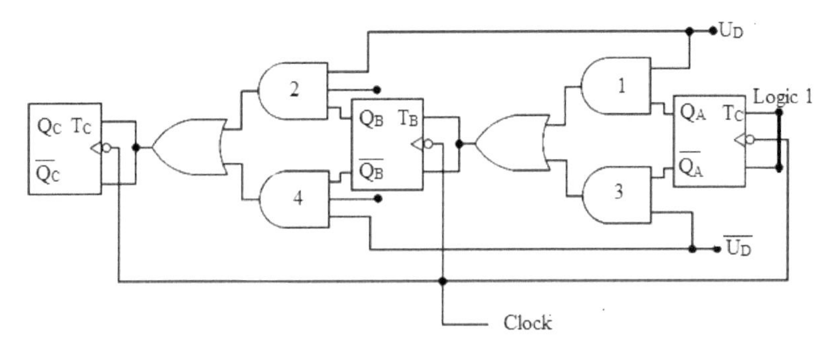

3-bit synchronous up/down counter using T flip-flop.

The figure shows the design of a 3-bit up/down counter which performs increment as well as decrement operation, using the same logic circuit.

- **Ques: 130. What are the main characteristics of RTL?**

Solution:
The characteristics of RTL are enlisted below:

1. Poor noise margin
2. Low speed of operation
3. High power dissipation
4. Poor fan-out

- **Ques: 131. What is the propagation delay in DTL gates?**

Solution:
The propagation delay is based on the switching of transistors. The propagation delay of diode-transistor logic (DTL) gates is typically in the range of 30–80 ns.

- **Ques: 132. What are the main factors that affect the propagation delay?**

Solution:

1. Storage time
2. RC time constant

Storage time is inversely proportional to the propagation delay. The RC time constant is directly proportional to the propagation delay.

- **Ques: 133. Name any three configurations in TTL gates.**

Solution:

1. Open collector output
2. Totem-pole output
3. Tri-state output

- **Ques: 134. Differentiate between Schottky TTL and fast TTL.**

Solution:

S. Nos.	Parameters	Schottky TTL	Fast TTL
1	Power dissipation	19 mW	4 mW
2	Propagation delay	3 ns	3 ns
3	Speed power product	57	12
4	Fan-out	10	20

- **Ques: 135. What is the noise margin?**

Solution:
During transmission and logic designing, unwanted signals interrupt the connecting devices, known as noise. Due to noise, the voltage level may raise or reduce. The ability of a circuit to tolerate the noise is known as noise immunity. The amount by which the circuit can tolerate the noise is called noise margin.

- **Ques: 136. What are the applications of the open-collector TTL gate?**

Solution:

1. Relay driving
2. Wired logic
3. Design a common bus system

- **Ques: 137. What are the output states of tri-state TTL?**

Solution:
As the name implies, there are three output states:

1. A logic 0 or low-level state, when the lower transistor in the totem-pole configuration is in ON mode and the upper transistor is in OFF mode.
2. A logic 1 or high-level state, when the lower transistor in the totem-pole configuration is in OFF mode and the upper transistor is in ON mode.
3. A third state, where both transistors in the totem-pole configuration are in OFF mode. This provides an open circuit or high impedance state which allows the direct-wired connection of various outputs on a common line.

- **Ques: 138. When the totem-pole output TTL gate is called a tri-state gate?**

Solution:
When the wired connections of outputs in totem pole TTL are interconnected to form a common bus system, then it acts as a tri-state TTL. This is a special type of totem pole TTL.

- **Ques: 139. Draw the graphical symbol of the tri-state buffer gate.**

Solution:

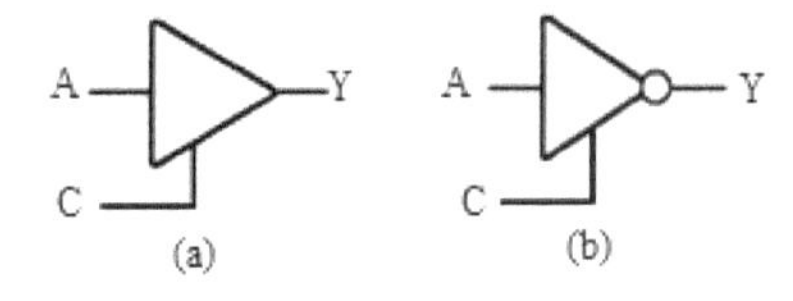

Tri-state buffer gate

- **Ques: 140. What is the noise margin of ECL circuits?**

Solution:
The noise margin of ECL circuits is less than TTL. The typical range of noise margin in ECL circuits varies from 0.2 to 0.25 V. ECL is not reliable for high-noise environments.

- **Ques: 141. What are the advantages and disadvantages of the CMOS logic family?**

Solution:
Advantages:

1. Low power consumption
2. Good fan-out
3. Low cost
4. Good noise-margin
5. Wide logic swings
6. Wide temperature range

Disadvantages

1. Slower than TTL
2. CMOS must be protected from static discharge, which can damage the silicon dioxide layer.

- **Ques: 142. Name the various families of CMOS.**

Solution:

1. CMOS 4000 series
2. 74 MC00 series
3. FACT (Fairchild Advanced CMOS Technology) series that has 74ACOO, 74FCTOO, and 74FCTAOO.

- **Ques: 143. How can CMOS be prevented from static discharge and transient voltages?**

Solution:

1. Store CMOS in static shield bags.
2. Battery-powered soldering irons should be used.
3. Input signals should not exceed power supply voltages.
4. Remove CMOS ICs only when power is off.
5. Connect unused leads with the ground.

- **Ques: 144. Give any three applications of CMOS ICs.**

Solution:

1. Calculators
2. Electronic wristwatches
3. Portable computers

- **Ques: 145. Differentiate between ECL and Schottky TTL.**

Solution:

S. Nos.	Parameters	Schottky TTL	ECL
1	Power dissipation	19 mW	60 mW
2	Propagation delay	3 ns	1 ns
3	Flip-flop clock frequency	125 MHz	500 MHz
4	Voltage swing	3.0 V	0.85 V

- **Ques: 146. What is the figure of merit? What are the units of the figure of merit?**

Solution:
The product of speed and power is known as a figure of merit.

$$\text{Figure of merit} = \text{propagation delay} \times \text{power dissipation}$$
$$= tp \times Pd = nsec \times mW = pJ$$

- **Ques: 147. What are fan-in and fan-out?**

Solution:
The fan-in is defined as the number of inputs that a logic gate can have. The gates with large fan-in are bigger and slower. The fan-out is the capability of a logic gate. It signifies the number of similar gates that a logic gate can drive. It is desirable to have a high fan-out. The gates with large fan-out are slower.

- **Ques: 148. Which logic family is having the lowest propagation delay?**

Solution:
The emitter-coupled logic (ECL) is having the lowest propagation delay as compared to other logic families. It makes ECL the fastest digital logic family.

- **Ques: 149. What is the advantage and disadvantage of DTL circuits?**

Solution:
DTL circuit has a better noise margin and great fan-out than RTL circuits. But DTL is having a slow speed of operation than TTL.

- **Ques: 150. What is the limitation of RTL?**

Solution:

1. Speed of operation is low
2. Fan out is less
3. Noise immunity is poor
4. Power dissipation is high
5. Temperature-sensitive

- **Ques: 151. What are pull-up and pull-down logic?**

Solution:
In CMOS, the PMOS region is connected to Vdd, i.e., high logic, so it is called pull-up and NMOS is connected to ground, i.e., low logic, so it is called pull-down.

- **Ques: 152. Draw CMOS NOR gate.**

Solution:

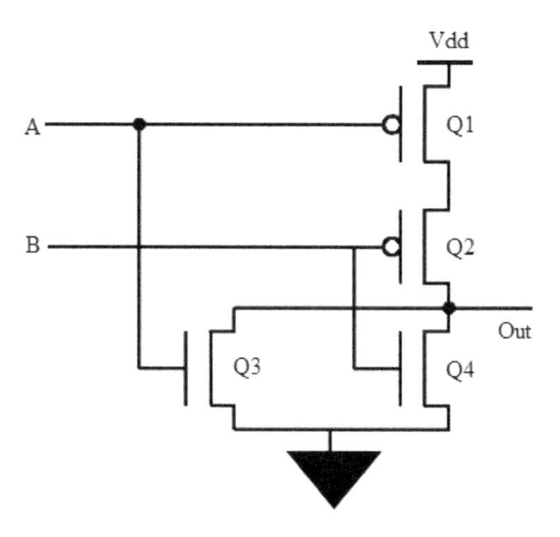

CMOS NOR

- **Ques: 153. Draw CMOS OR gate.**

Solution:

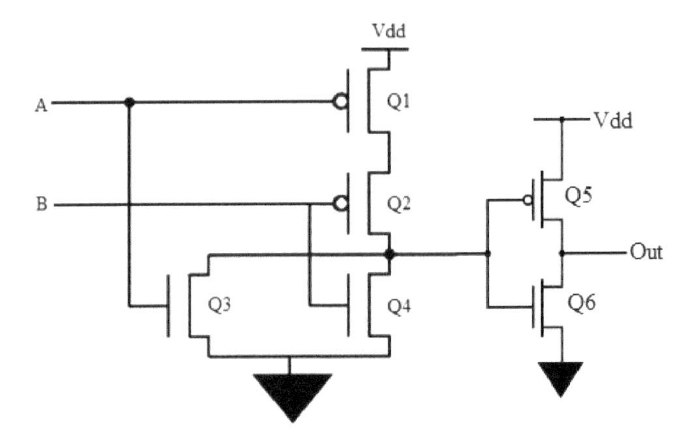

CMOS OR

- **Ques: 154. Draw CMOS AND gate.**

Solution:

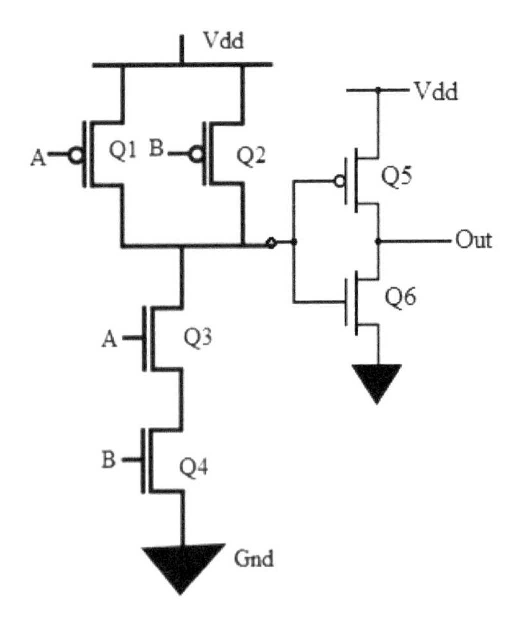

CMOS AND

- **Ques: 155. Design 65536 x 8 ROM using two 65536 x 4 ROM.**

Solution:

The number of address lines = 16 (as 2^{16} = 65536)

The address lines will remain the same, the data bus will increase. As per need, 8 bits data bus can be achieved by connecting two 4 bit data bus.

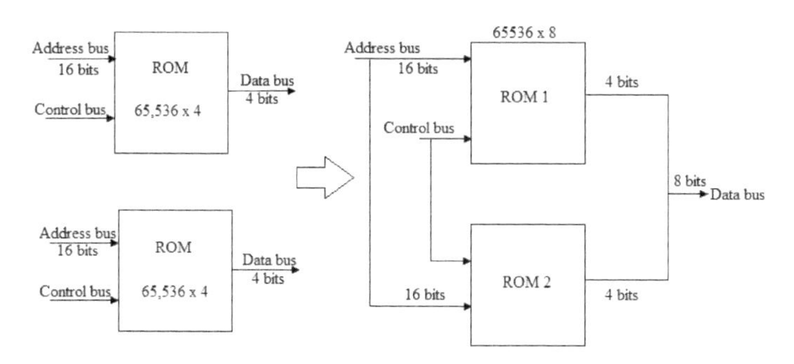

Word length expansion.

- **Ques: 156. Design 2M x 8 ROM using two 1M x 8 RAM cells.**

Solution:

1M memory = 1,048,576 addresses, it requires 20 address bits (as 1,048,576 = 2^{20})

2M memory = 2,097,152 addresses, it requires 21 address bits (as 2,097,152 = 2^{21})

During connection, the 21st address will determine which chip will be enable.

The figure shows the word capacity expansion, where two RAM are collaborated to form an expanded memory.

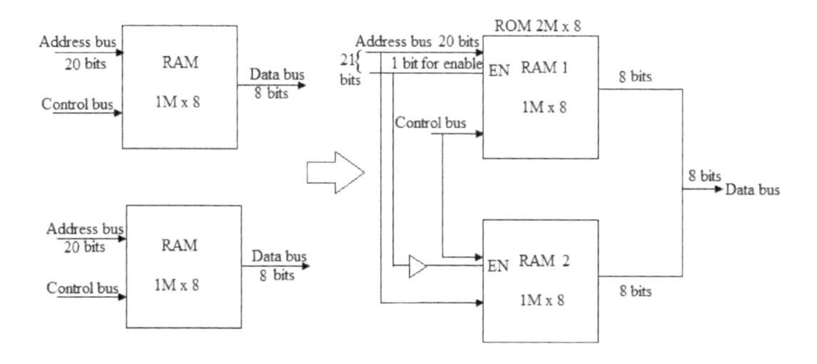

Word capacity expansion.

The extra address bit will select the ROM using the enable line. The particular RAM will be activated at a time for further operation. The extra bit is connected to both RAM through an inverter, which regulates the activation operation.

- **Ques: 157. Differentiate between EPROM and EEPROM.**

Solution:

Parameters	EPROM	EEPROM
Memory size	16M bit	1M bit
Chip size	$7.8 \times 17.39\ \mu m^2$	$11.8 \times 7.7\ \mu m^2$
Access time	62 nsec	120 nsec
Erase/Programming cycles	100	100000

- **Ques: 158. If 512K memory chip has 8 pins for data, then calculate (a) the number of locations and (b) number of address lines.**

Solution:
Memory capacity = 512K
Data bits = 8 (data bits = data pins)
(a) $Number\ of\ locations = \frac{chip\ capacity}{data\ bits} = \frac{512}{8} = 64\text{K}$
Memory organization = 64K × 8
(b) $Number\ of\ address\ lines = 16$
Because ($2^{16} = 64\text{K}$), therefore, address lines are 16.

- **Ques: 159. What are the commonly used memory chips?**

Solution:
The commonly used memory chips are the following:

1. RAM (random-access memory)
2. SAM (sequential-access memory)
3. ROM (read-only memory)

- **Ques: 160. What is rcead-only memory?**

Solution:
The read-only memory (ROM) is a memory unit, where data information is stored permanently. The data information remains stored in the device, whether the power is turned off and on again.

- **Ques: 161. Differentiate between volatile and nonvolatile memory.**

Solution:

Volatile memory	Nonvolatile memory
1. When the power is turned off, the data information is lost	1. When the power is switched off, the data is retained in the memory.
2. Examples are sctatic RAM and dynamic RAM	2. Examples are magnetic disk and ROM

- **Ques: 162. What do you mean by nondestructive readout?**

Solution:
During "read" operation, data information is read from the memory. In this way, the data is copied, but it is not destroyed from the memory. It retains in the memory as well. So, the process of copying the contents of memory location without destroying the contents is called nondestructive readout.

- **Ques: 163. Enlist any two types of semiconductor memories.**

Solution:

1. Random-access memory (RAM)
2. Read-only memory (ROM)

- **Ques: 164. What is the function of $read/\overline{write}$ input?**

Solution:
If $read/\overline{write}$ =1, the memory performs a read operation, i.e., information is retrieved from the memory location.
If $read/\overline{write}$ = 0, the memory performs a write operation, i.e., information is stored in the desired memory location.

- **Ques: 165. What is memory enable?**

Solution:
Memory enables or chip select input is used in the multichip environment. The function of enable input is to select the particular chip for the memory expansion process.

- **Ques: 166. Enlist the programming techniques of ROM.**

Solution:

1. Mask programming
2. Fuse programming
3. PROM
4. EPROM
5. EEPROM

- **Ques: 167. What are the various types of RAM?**

Solution:

The various types of RAM are thc following:

1. NMO RAM (nitride metal–oxide–semiconductor RAM)
2. CMOS RAM
3. Schottky TTL RAM
4. ECL RAM

- **Ques: 168. How the charge is measured in CCD IC?**

Solution:

An amplifier is connected with the search data register, which measures each charge and converts it into voltage. If the well is full, it signifies approximately 100,000 electrons, the conversion factor for the full well is 5–10 μV per electron.

- **Ques: 169. What is quantum efficiency in CCD IC?**

Solution:

It measures the percentage of photons that are detected by the light detector and converted into an electrical impulse.

- **Ques: 170. Which type of noise is present in thc CCD IC?**

Solution:

The sources of noise in CCD IC are the following:

1. Dark current
2. Readout noise

- **Ques: 171. Draw the block diagram of RAM.**

Solution:

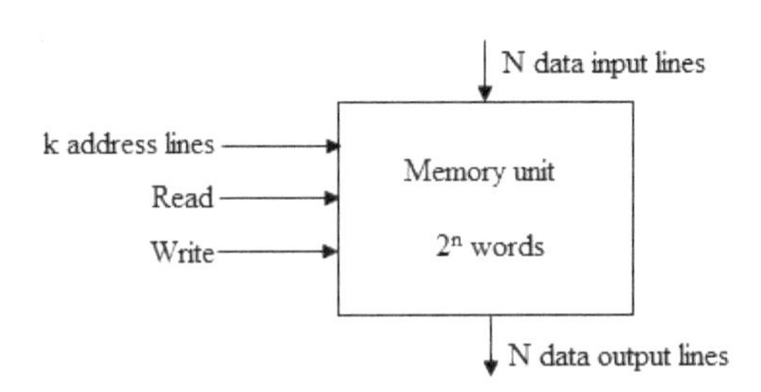

Block diagram of RAM.

- **Ques: 172. What is a cross-point in ROM?**

Solution:
The programmable intersection between two lines in ROM is known as cross-point. It acts like a switch that can be altered to either ON or OFF.

- **Ques: 173. What is the disadvantage of DRAM?**

Solution:
The disadvantage of DRAM is that the capacitor needs to be charged once every two milliseconds.

- **Ques: 174. What is MRAM?**

Solution:
Magneto-resistive random-access memory is abbreviated as MRAM. It is a nonvolatile memory that uses magnetic charges to store the data information. When the power is off, the data information in magneto-resistive RAM is not lost.

- **Ques: 175. What is the difference between CCD sensors and CMOS sensors?**

Solution:

Parameters	CCD	CMOS
Noise	Less susceptible to noise	More susceptible to noise
Light sensitivity	Light sensitivity is more	Light sensitivity is less
Power	Power consumption is high	Consumes less power
Cost	Fabrication cost is more	It can be fabricated on any silicon production line
Quality	Higher quality and more pixel	Average quality

- **Ques: 176. Which is better CMOS or CCD?**

Solution:
Charged coupled devices are of higher quality. The operation is faster than CMOS. Noise susceptibility is less in CCD as compared to CMOS. Regarding performance, quality, and durability, CCD is better than CMOS.

- **Ques: 177. What are the characteristics of DRAM?**

Solution:

1. The bits are stored in the form charge in the capacitor.
2. Charge leakage is present.
3. It is simple in construction.
4. It needs refreshing when powered.
5. Its cost is less per bit,
6. It acts as the main memory.
7. The speed of operation is slow.

- **Ques: 178. Enlist the main characteristics of SRAM.**

Solution:

1. Leakage charges are nil
2. More expensive
3. More complex in construction
4. The speed of operation is fast
5. Acts as cache memory
6. No need for refreshing when power is ON.
7. The bits are stored in the form of ON/OFF switches

- **Ques: 179. Differentiate between SRAM and DRAM.**

Solution:

Parameters	SRAM	DRAM
Capacity	Less memory cell per unit area	More memory cell per unit area
Access time	Less access time	Access time is more
Basic unit	Flip-flops	Capacitors
Cost	It is costly than SRAM	The cost is low
Use	Used for cache memory	Used for main memory

- **Ques: 180. How many address lines and data lines are needed in 4K $\times$ 16 memory?**

Solution:

$$
\begin{aligned}
1\text{K} &= 2^{10} = 1024 \\
4\text{K} &= 4(2^{10}) = 2^2(2^{10}) \\
4\text{K} \times 16 &= 2^2(2^{10}) \times 16 \\
&= 2^{12} \times 16
\end{aligned}
$$

Therefore, the number of address lines $= 12$

Number of data lines $= 16$

- **Ques: 181. How many address lines and data lines are needed in 2G × 16 memory?**

Solution:

$$
\begin{aligned}
1\text{G} &= 2^{30} \\
2\text{K} &= 2(2^{30}) = 2^1(2^{30}) \\
2\text{K} \times 16 &= 2^1(2^{30}) \times 16 \\
&= 2^{31} \times 16
\end{aligned}
$$

Therefore, the number of address lines = 31
Number of data lines = 16

- **Ques: 182. How many address lines and data lines are needed in 16M × 32 memory?**

Solution:

$$
\begin{aligned}
1\text{M} &= 2^{20} \\
16\text{M} &= 2^4(2^{20}) = 2^4(2^{20}) \\
16\text{M} \times 3 &= 2^4(2^{20}) \times 32 \\
&= 2^{24} \times 32
\end{aligned}
$$

Therefore, the number of address lines = 24
Number of data lines = 32

- **Ques: 183. What is the number of bytes stored in 4K × 16 memory?**

Solution:

$$
\begin{aligned}
1\text{ byte} &= 8\text{ bits} = 2^3 \\
4\text{K} &= 4(2^{10}) = 2^2(2^{10}) \\
4\text{K} \times 16 &= 2^{12} \times 2^4 \\
&= 2^{12} \times 2^1(2^3) \\
&= 2^{13} \times 2^3 \\
&= 2^{13}\text{ bytes}
\end{aligned}
$$

So, 2^{13} bytes stored in 4K × 16 memory

- **Ques: 184. What is the number of bytes stored in 16M × 32 memory?**

Solution:

$$\begin{aligned} 1 \text{ byte} &= 8 \text{ bits} = 2^3 \\ 16\text{M} &= 2^4(2^{20}) = 2^{24} \\ 16\text{M} \times 32 &= 2^{24} \times 2^5 \\ &= 2^{24} \times 2^2(2^3) \\ &= 2^{26} \times 2^3 \\ &= 2^{26} \text{ bytes} \end{aligned}$$

So, 2^{26} bytes stored in 16M $\times$ 32 memory

- **Ques: 185. What is the number of bytes stored in 2G $\times$ 16 memory?**

Solution:

$$\begin{aligned} 1\text{byte} &= 8 \text{ bits} = 2^3 \\ 2\text{G} &= 2(2^{30}) = 2^1(2^{30}) \\ 2\text{G} \times 16 &= 2^{31} \times 2^4 \\ &= 2^{31} \times 2^1(2^3) \\ &= 2^{32} \times 2^3 \\ &= 2^{32} \text{ bytes} \end{aligned}$$

So, 2^{32} bytes stored in 2G $\times$ 16 memory

- **Ques: 186. What are the parameters that determine the size of PLA?**

Solution:
The size of PLA is determined by

1. Number of inputs
2. Number of product terms
3. Number of outputs

- **Ques: 187. What are the steps to implement circuits with PLA?**

Solution:

1. Simplify Boolean functions to a minimum number of terms.
2. The true and complement of each function should be simplified to find which one can be expressed with fewer product terms and which one provides product terms that are common to other functions.

- **Ques: 188. Enlist the devices that are used to solve very complicated device problems?**

Solution:

The devices that can be used to handle complicated device problems are the following:

1. ROM (read-only memory)
2. PROM (programmable read-only memory)
3. PAL (programmable array logic)
4. Gate array
5. Programmable gate array

- **Ques: 189. Differentiate between PROM, PAL, and PLA.**

Solution:

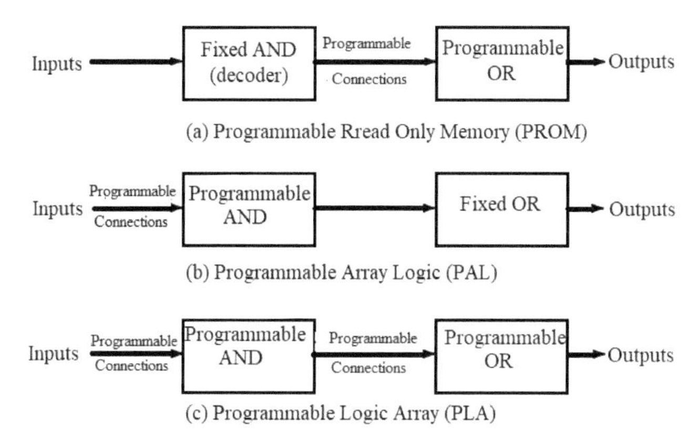

(a) Programmable Rread Only Memory (PROM)

(b) Programmable Array Logic (PAL)

(c) Programmable Logic Array (PLA)

- **Ques: 190. Differentiate between CPLD and FPGA.**

Solution:

S. Nos.	CPLD	FPGA
1	Complex programmable logic device is abbreviated for CPLD.	Field programmable gate array is abbreviated for FPGA.
2	It is a collection of PLDs	It is a collection of CLBs
3	The gate density is up to 10,000 gates	The gate density is 10^3–10^8 gates
4	Complex interconnections.	Programmable at the field.
5	AND-OR array in PAL-like blocks	LUTs are there in CLB
6	The configuration is stored in RAM	The configuration is stored in ROM
7	Configuration is nonvolatile	Configuration is volatile
8	Simpler architecture and few I/O registers.	FPGA has a high number of I/O registers.

- **Ques: 191. Differentiate between ASIC and FPGA.**

Solution:
ASIC is application specific integrated circuit, where an IC is designed specifically for the customer's need. It is different from FPGA in the following ways:

Parameters	FPGA	ASIC
Time to market	The time to market and design of FPGA is fast.	It is a slow process to market ASIC.
NRE	Nonrecurring expenses are low	Because of the cost of circuit design and mask, it has high nonrecurring expenses.
Design flow	Design flow is simple	Design flow is complex
Unit cost	Unit cost is high	Low
Performance	Its performance is medium	Its performance is high
Power consumption	The power consumption is high	The power consumption is low.
Unit cost	The unit cost of FPGA is medium	The unit cost of ASIC is low

- **Ques: 192. What do you mean by RTL?**

Solution:
RTL (register transfer level) design refers to the designing and modeling of digital circuits to produce synthesizable hardware models and design digital circuits at a higher level of abstraction using HDL. An RTL code can be tested on an FPGA board for its intended functionality by taking the .bit file format.

- **Ques: 193. What is the difference between simulation and synthesis?**

Solution:
Simulation is the process of using simulation software (simulator) to verify the functional correctness of a digital design that is modeled using an HDL (hardware description language) like Verilog/VHDL.

Synthesis is a process in which a design behavior that is modeled using an HDL is translated into an implementation consisting of logic gates. This is done by a synthesis tool which is another software program. Synthesis converts your code to a net list. With the help of various tools, the net list is mapped to hardware.

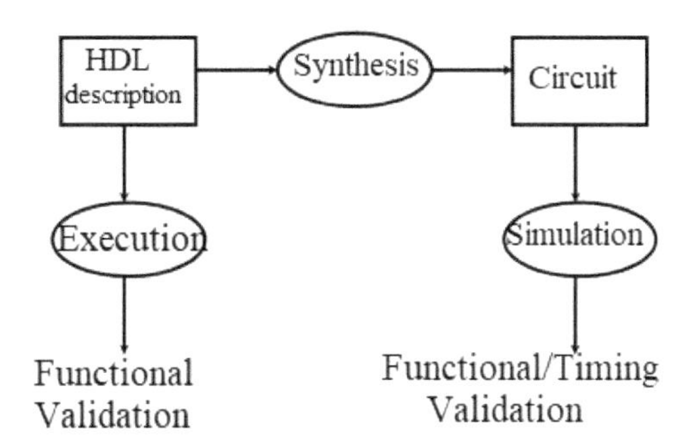

- **Ques: 194. Differentiate between verification and validation.**

Solution:

Parameters	**Verification**	**Validation**
Objective	It ensures whether the product is built as per design specifications.	It ensures that the product meets the user's specifications
Process	Whether the output is as per input or not.	Whether the software is accepted by the user or not.
Outcome	Are we building a product, right? The software should conform to its specifications.	Are we building the right product? The software should do what the user requires.
Evaluated items	Plans, specifications, design specifications, code, test cases.	Actual product
Testing	Static	Dynamic
Order	Verification is done before validation.	Validation is done after verification.
Code	It does not involve executing the code.	It involves the execution of code.

The verification and validation can be shown in the following figure.

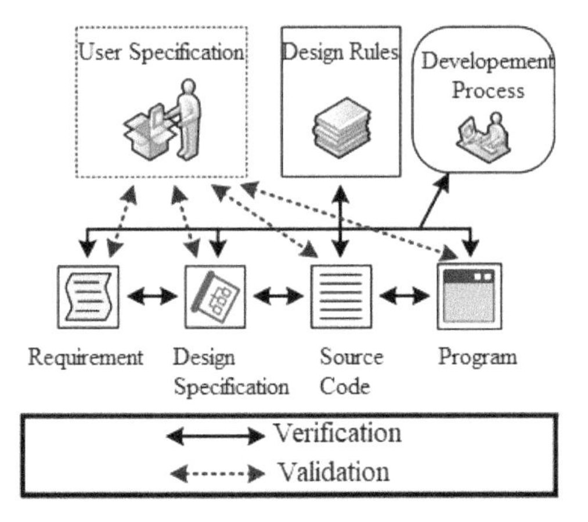

- **Ques: 195. What are the advantages of PLDs?**

Solution:

1. The cost is less.
2. Low power requirements
3. Highly reliable
4. Design software is available

- **Ques: 196. What are the applications of FPGA?**

Solution:
The major application areas of FPGA are the following:

1. Emulation of entire hardware, viz. FPGA
2. Custom computing machines.
3. FPGA is used in device controller, encoding, and filtering.
4. FPGA provides a solution for machine vision, industrial automation, and video surveillance.
5. Bioinformatics and ASIC prototyping.

- **Ques: 197. Enlist any two applications of CPLD.**

Solution:
Due to low cost and short delay, CPLDs are used in

1. Address decoding
2. Cost-sensitive battery-operated portable applications

- **Ques: 198. Enlist the main advantages of PLDs.**

Solution:

1. Flexible design
2. Improved reliability
3. Reduction in power consumption
4. Complexity is reduced
5. Field programmable
6. Erasable and programmable

- **Ques: 199. What are the main applications of PLDs?**

Solution:

1. Glue logic
2. Synchronization
3. Bus interface
4. Decoder
5. Counters
6. State machines

- **Ques: 200. Design an exclusive-NOR gate using FPGA with 2-input LUT.**

Solution:
Using Shannon's expansion, the designing of the logic function using LUT can be done.

The truth table of the exclusive-NOR gate is

X1	X2	F
0	0	1
0	1	0
1	0	0
1	1	1

$F = X1X2 + \overline{X1}.\overline{X2}$

Or using Shannon's expansion

$$F = \overline{X1}(\overline{X2}) + X1X2$$
$$F = \overline{X1}(\overline{X2}(1) + X2(0)) + X1(\overline{X2}(0) + X2(1))$$

It can be shown in the following figure.

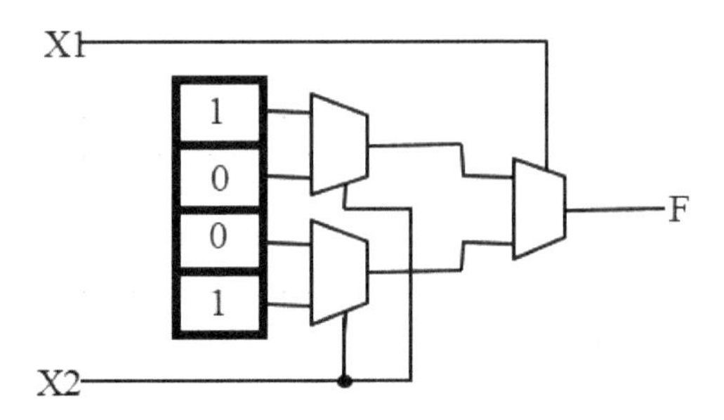

2 Input LUT design for Ex-NOR.

Similarly, all logic functions can be implemented using Shannon's expansion or Boole's expansion theorem.

- **Ques: 201 What is the difference between analog and digital signals?**

Solution:

S. Nos.	Analog signals	Digital signals
1	It is a continuous signal that changes over some time	It has a discrete nature at each sampling point.
2	The data information is in the form of a continuous wave.	The data information is in the form of binary bits.
3	Analog signals are represented by a sign wave.	Digital signals are represented by a square wave.
4	It is characterized by amplitude, frequency, or phase.	It is characterized by a bit rate or baud rate.
5	It is more prone to noise and distortion.	It is less prone to noise and distortion.
6	There is no fixed range of analog signals.	The digital signals are represented either in 0 or 1.
7	An example of an analog signal is the human voice, old radio sets, old telephone sets, etc.	The example of a digital signal is CD, DVD, computer, and digital devices.

- **Ques: 202. What is the need for converters?**

Solution:
In the real-world, the signals are analog. These signals are continuous. But the processing of signals in the digital environment is much easier, faster, efficient, and less distorted by noise. So, conversion from analog to digital is needed. However, the output is again required in analog form. That's why digital to analog conversion is required.

- **Ques: 203. What are the two main factors on which the accuracy of ADC rely?**

Solution:

1. The sampling rate
2. The resolution

With an increase in sampling rate and resolution, the accuracy of ADC can be increased.

- **Ques: 204. Explain the Nyquist–Shannon theorem of sampling.**

Solution:
The minimum sampling rate should be at least twice the highest data frequency of the analog signal, i.e., $fs > 2 \times f(max)$

- **Ques: 205. In the 8-bit converter, if the sampling rate is 1 ms and full-range scale is 5 V then find**

(a) number of possible states
(b) resolution
Solution:
(a) No. of possible states $= 2^{n-1} = 2^8 - 1 = 255$
(b) Resolution $= \frac{Vfs}{N} = \frac{5}{8} = 19.5$ mV

- **Ques: 206. What is aliasing? When it occurs?**

Solution:
Aliasing occurs when you sample a signal (anything which repeats a cycle over time) too slowly (at a frequency comparable to or smaller than the signal being measured), and obtain an incorrect frequency and/or amplitude as a result. It occurs when the Nyquist–Shannon's theorem violates.

- **Ques: 207. What do you mean by settling time in D/A converters?**

Solution:
The settling time represents the time it takes for the output to settle within a specified band ± (1/2) LSB of its final value following a code change at

the input (usually a full-scale change). It depends on the switching time of the logic circuitry due to internal parasitic capacitances and inductances. A typical settling time ranges from 100 ns to 10 μs depending on the word length and type of circuit used.

- **Ques: 208. What do you understand by stability in D/A converters?**

Solution:
The ability of a D/A converter to produce a stable output all the time is called stability. The performance of a converter changes with drift in temperature, aging, and power supply variations. So, all the parameters such as offset, gain, linearity error and monotonicity may change from the values specified in the datasheet. Temperature sensitivity defines the stability of a D/A converter.

- **Ques: 209. What are the advantages of dual slope A/D converter?**

Solution:
The dual-slope type A/D converter is insensitive to errors in the component values. The accuracy level is high.

- **Ques: 210. What do you mean by the term repeatability in A/D converters?**

Solution:
The repeatability is the specification of the A/D converter which depicts how well the output repeats the input, i.e., during conversion, how accurate the A/D converter generates the same data. It depends on the fact that how efficiently and effectively A/D converter processes and converts the data. If repeatability is ensured for the circuit, the accuracy will be automatically increased. For better repeatability, the distortion or surrounding noise should be filtered or reduced well.

- **Ques: 211. What is the disadvantage of a successive approximation A/D converter?**

Solution:

1. The speed is limited, i.e., 5 Mbps
2. The resolution is low.

- **Ques: 212. How do you measure resolution in ADC or DAC?**

Solution:
The resolution of D/A converter is the smallest change in output, which may be produced at the output or input terminal of the converter. It signifies how

finely the output may change between discrete steps.

$$Resolution = \frac{Vfs}{2^n - 1}$$

Here, n is the number of bits and V_{fs} is full-scale output.

- **Ques: 213. What do you mean by charge injection in A/D converters?**

Solution:
The A/D converter that uses a track/hold buffer must periodically connect its track/hold capacitor to the input signal, an action that causes a small inrush current. The charge injection is the characteristic of the track/hold capacitor requires time to charge to the correct voltage.

- **Ques: 214. Enlist any two smart A/D converters.**

Solution:

1. The MAX7651 combines a microcontroller with a 12-bit ADC.
2. The MAX1460 is a smart, sensor signal conditioner that implements a nonlinear compensation equation.

- **Ques: 215. What is the principal for algorithmic A/D converters?**

Solution:
The algorithmic A/D converters are based on comparisons between the reference voltage and the analog input signal. In this architecture, the reference is constant. The input signal changes in the range of zero to the reference voltage. In this method, the input signal is doubled and compared to the reference. As long as the input signal is less than the reference, the output is set to zero. The input signal is multiplied by 2 and is compared to the reference until it exceeds the reference. Then the output is set to 1 and the reference is subtracted from the input signal. The process can be repeated to achieve higher resolution.

- **Ques: 216. In which conditions the algorithmic A/D converters are used?**

Solution:
An algorithmic ADC is used in high-resolution and low-power applications.

- **Ques: 217. Which A/D converter is called parallel encoder?**

Solution:
Flash type A/D converter is called a parallel encoder. It is used for high speed and low resolution-based applications.

- **Ques: 218. In an 8-bit D/A converter, if the input is 00110010 then what is the maximum output voltage that produces 1.0 V.**

Solution:

$$Analog\ voltage = Proportionality\ factor\ (k) \times digital\ voltage$$
$$(00110010)2 = (50)10$$
$$1V = K \times 50$$
$$K = 20\ \mathrm{mV}$$

The largest (maximum) voltage will be for input $11111111 = (255)_{10}$.

$$Vout(\mathrm{max.}) = 20\mathrm{mv} \times 255$$
$$Vout(\mathrm{max.}) = 5.10\ \mathrm{V}$$

- **Ques: 219. A 5-bit D/A converter has a current output. For a digital input of 101000, an output current of 10 mA is produced. What will IOUT be for a digital input of 11101?**

Solution:
$(10100)2 = (20)10.$
Since IOUT = 10 mA
Proportionality factor (k) = 0.5 mA. (K is step size/proportionality factor)
For $(11101)_2 = (29)_{10}$
IOUT = (0.5mA) $\times$ 29
= 14.5 mA

- **Ques: 220. A 10-bit D/A converter has a step size of 10 mV. Determine the full-scale output voltage and the percentage resolution.**

Solution:
As the number of bits in D/A converter $= 10$
The number of steps $= 2^{10} - 1 = 1023$ of 10 mV
Full scale output $= 10\ mV \times 1023 = 10.23$ V
% Resolution $= \frac{10\mathrm{mV}}{10.23\mathrm{V}} \times 100 = 0.1\%$

- **Ques: 221. Differentiate between VHDL and Verilog.**

Solution:
VHDL is an abbreviated form of VHSIC HDL, i.e., very high-speed integrated circuit hardware description language. Both VHDL and Verilog are

HDLs. The differences in both languages are enlisted in the following table:

S. Nos.	VHDL	Verilog
1	The compilation order does not affect the simulation.	Proper care should be taken for compilation order.
2	User-defined data types are allowed.	User-defined data types are not allowed.
3	It is like ADA and hard to understand.	Verilog is like C language and it is easy to learn.
4	The IEEE VHDL library is present.	There is no concept of the library.
5	In VHDL packages, procedural and statements are present.	There is no concept of packages.
6	Unary reduction operator is not present.	Unary reduction operator is present.
7	Concurrent procedural calls are allowed	Concurrent procedural calls are not allowed
8	VHDL supports multidimensional array.	Verilog does not support multidimensional array.
9	Number of instances can be replicate using generate a statement	There is no concept of generating a statement
10	User-defined primitives cannot be designed in VHDL.	Primitives can be designed as per user requirements.
11	There is no concept of switch level modeling. Only structural, data flow, and behavioral modeling styles are present in VHDL.	There are four types of modeling, i.e., structural, data flow, behavioral, and switch level modeling.

- **Ques: 222. What is the difference between $display, $monitor, and $strobe?**

Solution:

$display and $strobe display once every time they are executed, whereas $monitor displays every time one of its parameters changes. The difference between $display and $strobe is that $strobe displays the parameters at the very end of the current simulation time unit rather than exactly where it is executed.

- **Ques: 223. What is the sensitivity list?**

Solution:
The sensitivity list indicates that when a change occurs to any one of the elements in the list change, begin-end statement inside that always blocks will get executed.

- **Ques: 224. Write a Verilog code for a synchronous and asynchronous reset.**

Solution:
In synchronous reset, the clock must be present in the sensitivity list.

```
always @ (posedge clk)
begin
if (reset)
    -
    -
    end
```

In the asynchronous reset, the reset must be present in the sensitivity list.

```
    always @(posedge clock or posedge reset)
begin
if (reset)
-
    -
    end
```

- **Ques: 225. Write a Verilog code to swap the contents of two registers with and without a temporary register.**

Solution:
The contents can be swapped either by using a temporary register or by using nonblocking statements.

With temp register

```
always @ (posedge clock)
begin
temp=b;
b=a;
a=temp;
end
```

Without temp register (nonblocking)

```
always @ (posedge clock)
begin
a <= b;
b <= a;
end
```

- **Ques: 226. What is the difference between the following two lines of Verilog code?**
  1. #10 a = b;
  2. a = #10 b;

Solution:

In the first line of code, #10 a = b; it is a delay of 10 ns (time units) before performing a = b;

In the second line of code,

a = #10 b; the value of b is calculated and stored in an internal temp register.

After 10 time units, assign this stored value to a.

- **Ques: 227. What do you mean by $fileopenr, $fileopenw, and $fileopena?**

Solution:

The function $fopenr opens an existing file for reading. $fopenw opens a new file for writing, and $fopena opens a new file for writing where any data will be appended to the end of the file.

- **Ques: 228. What is the difference between $reset, $stop, and $finish?**

Solution:

$reset resets the simulation back to time 0; $stop halts the simulator and puts it in interactive mode where the user can enter commands; $finish exits the simulator back to the operating system.

- **Ques: 229. What do you mean by PLI?**

Solution:

Programming language interface (PLI) of Verilog HDL is a mechanism to interface Verilog programs with programs written in C language. It also provides a mechanism to access internal databases of the simulator from the C program.

- **Ques: 230. If you have two ex-or gates, how will you do connections such that one will act as buffer and others will act as inverter?**

Solution:
When one input of the EX-OR gate is connected to logic 1 (VDD), it will act as an inverter.
When one input of the EX-OR gate is connected to logic 0 (GND), it will act as an inverter.

- **Ques: 231. What do you mean by clock skew? What is the reason for the clock skew? How can you minimize it?**

Solution:
In circuit design, clock skew is a phenomenon in synchronous circuits in which the clock signal arrives at different components at different times.

This is typically due to two causes.

1. Material flaw
2. Circuit size

Due to material property, signal either travel at abnormal rate and signal arrive at different components at different times. Clock skew can be minimized by the proper routing of the clock signal or putting variable delay buffer so that all clock inputs arrive at the same time.

- **Ques: 232. What do you mean by slack? How it is measured?**

Solution:
In a circuit, sometimes delay occurs, which causes the event to be held at some other time rather than deadline/actual time. It is measured as the difference between actual time and operational time.

$$Slack = Actual\ time - operational\ time$$

- **Ques: 233. What do you mean by glitch?**

Solution:
Glitch is a temporary false or unwanted output. It is a kind of short duration pulse, mostly because of cross-talk from a neighbor clocking or switching circuits or sudden turn on/off in the circuit of interest.

The reasons for glitch maybe the following:

1. Unanticipated power supply noise
2. Inductive and capacitive cross-talk
3. Cosmic ray decay events due to the fact the passage of an energized particle leaves a trail of electrical charge behind it.

- **Ques: 234. Implement AND Gate using 2:1 MUX.**

Solution:

A two-input AND gate has inputs A and B. The 2:1 multiplexer requires one select line. So, "A" is fixed as a select line. The input value of 2:1 multiplexer can be calculated by exploring the relation between "B" and "Y," at different values of "A."

The truth table of AND gate

Input		Output	
A	B	Y	
0	0	0	0
0	1	0	
1	0	0	B
1	1	1	

This can be realized using 2:1 multiplexer as

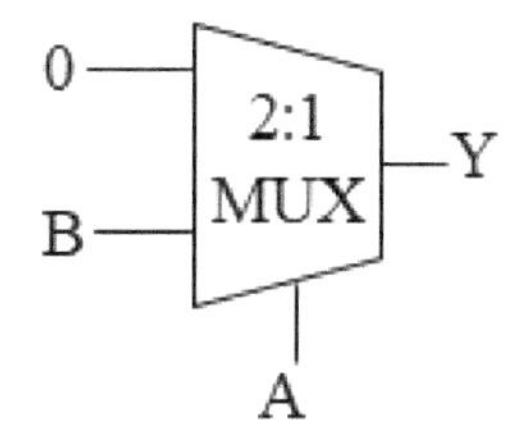

AND gate using 2:1 MUX

- **Ques: 235. Implement OR gate using 2:1 MUX.**

A two-input OR gate has inputs A and B. The 2:1 multiplexer requires one select line. So, "A" is fixed as a select line. The input value of 2:1 multiplexer can be calculated by exploring the relation between "B" and "Y," at different values of "A."

The truth table of OR gate

Input		Output	
A	B	Y	
0	0	0	B
0	1	1	
1	0	1	1
1	1	1	

This can be realized using 2:1 multiplexer as

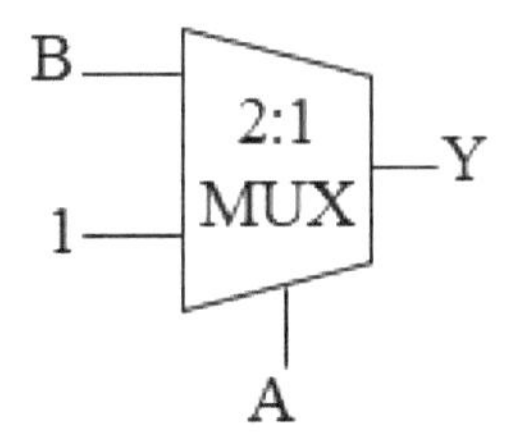

OR gate using 2:1 MUX.

- **Ques: 236. Implement NAND gate using 2:1 MUX.**

Solution:

A two-input NAND gate has inputs A and B. The 2:1 multiplexer requires one select line. So, "A" is fixed as a select line. The input value of 2:1 multiplexer can be calculated by exploring the relation between "B" and "Y," at different values of "A."

The truth table of NAND gate

Input		Output	
A	B	Y	
0	0	1	1
0	1	1	
1	0	1	B'
1	1	0	

This can be realized using 2:1 multiplexer as

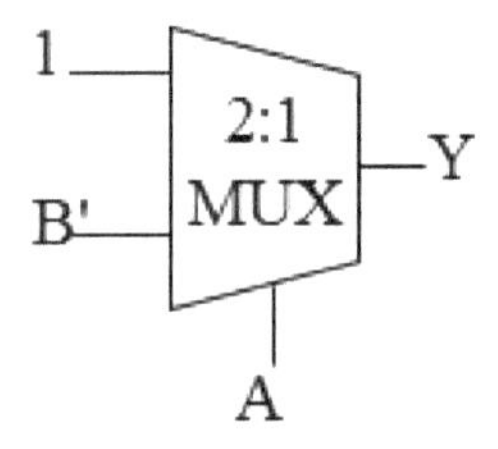

NAND gate using 2:1 MUX.

- **Ques: 237. Implement NOR gate using 2:1 MUX.**

Solution:

A two-input NOR gate has inputs A and B. The 2:1 multiplexer requires one select line. So, "A" is fixed as a select line. The input value of 2:1 multiplexer can be calculated by exploring the relation between "B" and "Y," at different values of "A."

The truth table of NOR gate

Input		Output	
A	B	Y	
0	0	1	B'
0	1	0	
1	0	0	0
1	1	0	

This can be realized using 2:1 multiplexer as

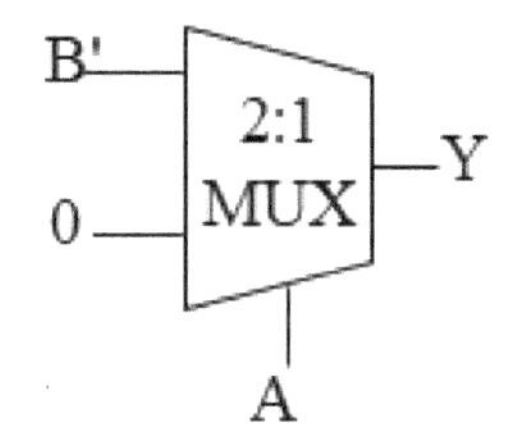

NOR gate using 2:1 MUX.

- **Ques: 238. Implement exclusive-OR gate using 2:1 MUX.**

Solution:

A two-input XOR gate has inputs A and B. The 2:1 multiplexer requires one select line. So, "A" is fixed as a select line. The input value of 2:1 multiplexer can be calculated by exploring the relation between "B" and "Y," at different values of "A."

The truth table of XOR gate

Input		Output	
A	B	Y	
0	0	1	B'
0	1	0	
1	0	0	B
1	1	1	

This can be realized using 2:1 multiplexer as

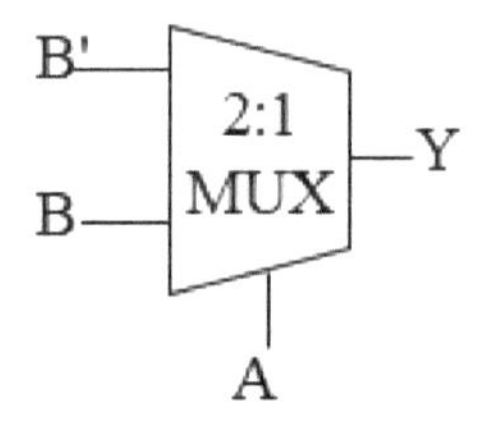

XOR gate using 2:1 MUX.

- **Ques: 239. Implement exclusive-NOR gate using 2:1 MUX.**

A two-input X-NOR gate has inputs A and B. The 2:1 multiplexer requires one select line. So, "A" is fixed as a select line. The input value of 2:1 multiplexer can be calculated by exploring the relation between "B" and "Y," at different values of "A."

The truth table of XNOR gate

Input		Output	
A	B	Y	
0	0	0	B
0	1	1	
1	0	1	B'
1	1	0	

This can be realized using 2:1 multiplexer as

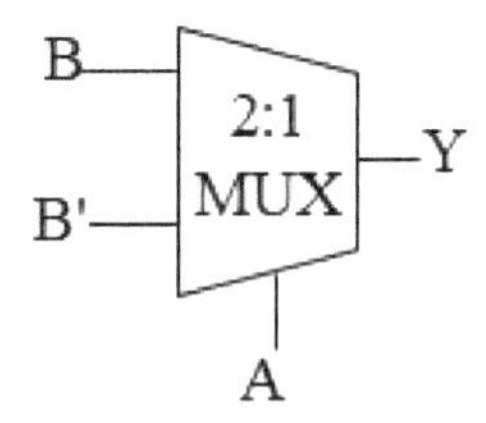

X-NOR gate using 2:1 MUX.

- **Ques: 240. Implement NOT gate using 2:1 MUX.**

Solution:

A two-input NOT gate has inputs A and B. The 2:1 multiplexer requires one select line. So, "A" is fixed as a select line. The input value of 2:1 multiplexer can be calculated by exploring the relation between "B" and "Y," at different values of "A."

The truth table of NOT gate

Input	Output
A	Y
0	1
1	0

This can be realized using 2:1 multiplexer as

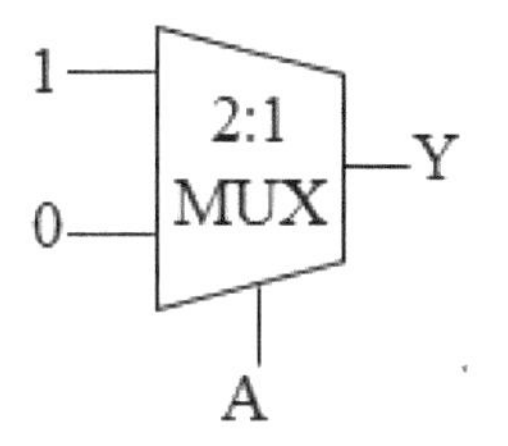

NOT gate using 2:1 MUX.

- **Ques: 241. Write the IC numbers of all digital components.**

Solution:

S. Nos.	Digital Logic	Parameters	IC/Board Numbers
1	Logic gates	Quad 2-input AND logic gate	7408
2		Quad 2-input OR logic gate	7432
3		NOT logic gate/hex inverter	7404
4		Quad 2-input NAND logic gate	7400
5		Quad 2-input NOR logic gate	7402
6		Quad 2-input Exclusive OR logic gate	7486
7		Quad 2-input Exclusive NOR logic gate	74266 (TTL) 4077 (CMOS)
8	Multiplexer	2:1 Multiplexer	74157

S. Nos.	Digital Logic	Parameters	IC/Board Numbers
9		4:1 Multiplexer	74153
10		8:1 Multiplexer	74151
11		16:1 Multiplexer	74150
12	Demultiplexer	1:2 Demultiplexer	74LVC1G19
13		1:4 Demultiplexer	74139
14		1:8 Demultiplexer	74138
15		1:16 Demultiplexer	74154
16	Decoder	2: 4 Decoder	74155 (TTL)
17		3:8 Decoder	74137/74138
18		4:16 Decoder	74154
19		BCD to decimal decoder	7441
20		BCD to seven segment decoders	7446/7447
21	Encoder	8:3 Priority Encoder	74148
22		10:4 Priority Encoder	74147
23	Digital Comparator	4-bit magnitude Comparator	7485
24		8-bit magnitude Comparator	74682
25	Flip-flop	SR Flip-flop	74279
26		JK Flip-flop	7470
27		JK Master Slave Flip-flop	7471
28		D Flip-flop	7474/7479
29		T Flip-flop	7473 short J and K
30	Shift register	8-bit Serial-In-Serial-Out register (SISO)	7491
31		8-bit Serial-In-Parallel-Out register (SIPO)	74164

S. Nos.	Digital Logic	Parameters	IC/Board Numbers
32		16-bit Parallel-in-Serial-Out register (PISO)	74674
33		4-bit Parallel-in-Parallel-Out register (PIPO)	7495
34	ADC and DAC	16-bit A/D converter	ADS5482 (TI)
35		16- bit D/A converter	DAC8728 (TI)
36	Adder & Subtractor	2-bit Full Adder	7482
37		4-bit Full Adder	7483
38		4-bit Full Subtractor	74385
39	Counter	Up-down binary counter	74191
40		Up-down decade counter	74190
41		Modulo 10 counters	74416
42	Programmable Logic Devices (PLD)	Field Programmable Gate Array (FPGA)	SPARTAN 6 family, ARTIX 7 family
43		Complex Programmable Logic Device (CPLD)	ALTERA MAX 7000 series
44	Memories	16-bit RAM	7481/7484
45		64-bit RAM	7489
46		256-bit ROM	7488
47		512-bit ROM	74186 (open collector)
48		256-bit PROM with open collector output	74188

S. Nos.	Digital Logic	Parameters	IC/Board Numbers
49		2048-bit PROM with open collector output	74470
50		2048-bit PROM with three-state output	74471
51		1024-bit PROM with three-state output	74287

- **Ques: 242. Differentiate MUX, DEMUX, encoder, and decoder in terms of the number of inputs and outputs.**

Solution:

Variables	Number of Inputs	Number of Outputs
MUX	2^N	1
DEMUX	1	2^N
ENCODER	2^N	N
DECODER	N	2^N

- **Ques: 243. How will you design a 4-Input NAND gate using only a 2-Input NAND gate?**

Solution:
Connect two inputs of the NAND gate to get an inverter.

- **Ques: 244. How many unused states are there in the 3-bit Johnson counter?**

Solution:

$$Number\ of\ unused\ states = 2^n - 2n = 2^3 - 2(3) = 8 - 6 = 2$$

The number of unused states = 2.

- **Ques: 245. Assume that you have two processors of similar configuration. One has clock skew 200 ps and the other has 400 ps. Which processor has more power?**

Solution:
Clock skew of 200 ps is more likely to have clock power.
Clock skew is inversely proportional to clock tree design and several buffers/overheads.

- **Ques: 246. What is the advantage and disadvantage of asynchronous reset?**

Solution:
Advantages:

1. The reset will not be added to the data path. This is known as reset release or reset removal.
2. No need for the clock to reset the circuit.

Disadvantage:
There is the possibility of metastability if the release of the reset occurred on or near a clock edge.

- **Ques: 247. Can you reduce clock skew to zero?**

Solution:
Although theoretically, we can reduce clock skew by using clock layout strategies, i.e., H tree, practically the process variations in R and C across the chip will consume excessive area and cause clock skew.

- **Ques: 248. How will you differentiate FIFO from RAM?**

Solution:

1. FIFO is used for synchronization purpose, whereas RAM is for storage purposes.
2. FIFO does not have address lines but RAM has address lines.

- **Ques: 249. What do you mean by race around the condition and how it can be eliminated?**

Solution:
When the clock signal is high and J = K = 1, the output keeps on toggling until the clock pulse is low. This is called the race around the condition. It can be eliminated by introducing Master-Slave JK Flip-flop, in which clock pulse has a short time duration than the propagation delay time of flip-flop.

- **Ques: 250. What do you mean by RAS and CAS?**

Solution:
RAS: Row Address Strobe: During SDRAM operation, the first address word is latched into the DRAM chip with RAS to save input pins.
CAS: Column Address Strobe: The second address word is latched with CAS. After RAS and CAS, the data is processed for a read operation.

- **Ques: 251. Give any two applications of buffers.**

Solution:

1. For small delays, buffers are used.
2. For providing high fan-out
3. Due to close routing, interelectrode capacitances cause cross-talk. It can be removed using a buffer.

- **Ques: 252. What do you understand by cross-talk?**

Solution:

During signal transmission, due to undesired capacitive or inductive coupling within a circuit or channel, signal creates an undesired effect on another channel or circuit. This is known as cross-talk. Its practical example is hearing disturbance during voice conversation via telephone.

- **Ques: 253. If in a D flip-flop, the clock and D input is shorted or tied together, what will happen in the circuit? Will it work as required?**

Solution:

If D input and clock are connected, the circuit will go to an indeterministic state known as the metastable state for an unknown amount of time. As, in such a situation, with rising of the clock the data will also rise or disrupt, it will violate the setup and hold time constraints.

- **Ques: 254. How can you find the fan-out limit of a circuit?**

Solution:

Fan-out is the ratio of load capacitance to input gate capacitance or in simple words you can say, it is the ratio of driven gate size to the driving gate size.

- **Ques: 255. What is the reason that in Karnaugh map the sequence is 00 01 11 10 rather than 00 01 10 11?**

Or

Why the K-map does not have terms in ordered numerical sequence?

Solution:

As you see, the preferred order of K-map is 00 01 11 10, in each sequence, there is a change of only 1 bit. For example, 00 to 01, only one bit is changed and so on.

The reasons for the unordered numerical sequence are the following:

1. It is done intentionally that crossing each horizontal or vertical cell boundary will reflect one variable change.

2. It is based on gray code, so the sequence is 00 01 11 10. The gray codes are used in K-map implementation because it reduces redundancy.
3. Asynchronous cannot take two values at the same time. So, the sequence is 00 01 11 10.
4. Grey codes follow adjacency property, i.e., between two successive gray codes there is only one-bit change permitted, so the sequence is 00 01 11 10.
5. In Boolean law, the equation gets simplified, only if there is a change in one bit. For example, if A is 0 and $\overline{A} = 1$, then $\overline{A} + A = 1$, etc., otherwise the output remains intact to input, 0 + 0 = 0, 1 + 1 = 1. So, the change of one variable is a must.

- **Ques: 256. How does the Boolean logic control the logic gates?**

Solution:
In digital electronics, the presence of voltage is denoted by high logic, i.e., logic 1 and ground logic is represented by logic 0, i.e., low logic. In simple words, 1 represents the ON state and 0 represents the OFF state.

- **Ques: 257. How many types of adders are there?**

Solution:
Adders are one of the most widely digital components in the digital IC design and are the necessary part of digital signal processing (DSP) applications. With the advances in technology, researchers have tried and are trying to design adders that offer either high speed, low power consumption, less area, or a combination of them. The types of adder are the following:

1. Ripple carry adder
2. Carry skip adder
3. Carry increment adder
4. Carry look ahead adder
5. Carry save adder
6. Carry select adder
7. Carry bypass adder

- **Ques: 258. What do you mean by a ripple carry adder?**

Solution:
A ripple carries adder is a logic circuit in which the carry-out of each full adder is the carry-in of the succeeding next most significant full adder. It is called a ripple carry adder because each carry bit gets rippled into the next stage.

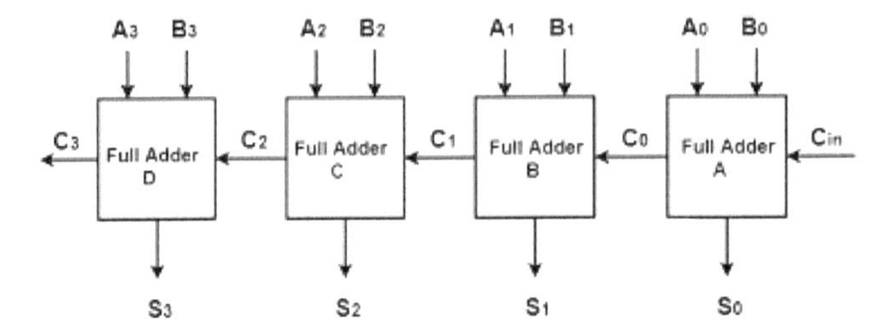

- **Ques: 259. What do you mean by carry skip adder?**

Solution:

A carry-skip adder (also known as a carry-bypass adder) is an adder implementation that improves on the delay of a ripple-carry adder with little effort compared to other adders. The improvement of the worst-case delay is achieved by using several carry-skip adders to form a block-carry-skip adder.

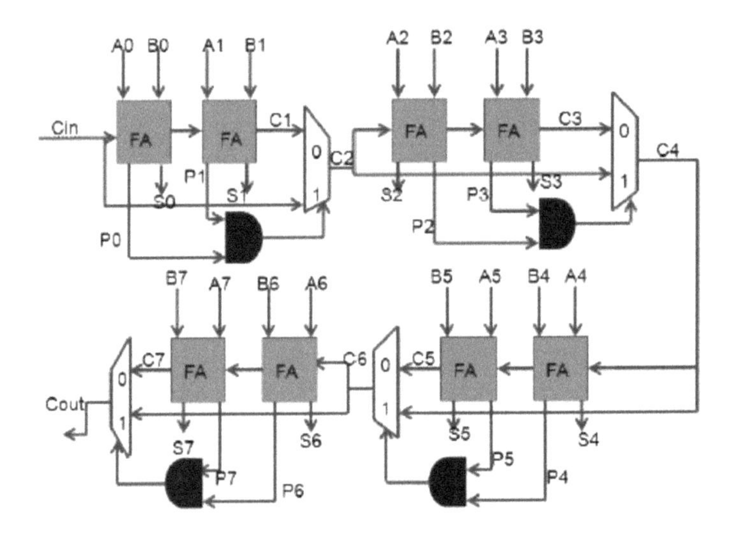

- **Ques: 260. What do you mean by carry increment adder?**

Solution:

The design of carry increment adder (CIA) consists of RCAs and incremental circuitry. The incremental circuit can be designed using HAs in a ripple carry chain with a sequential order. The addition operation is done by dividing a total number of bits into a group of 4 bits and addition operation is done using several 4-bit RCAs. The architecture of the CIA is shown in the following figure:

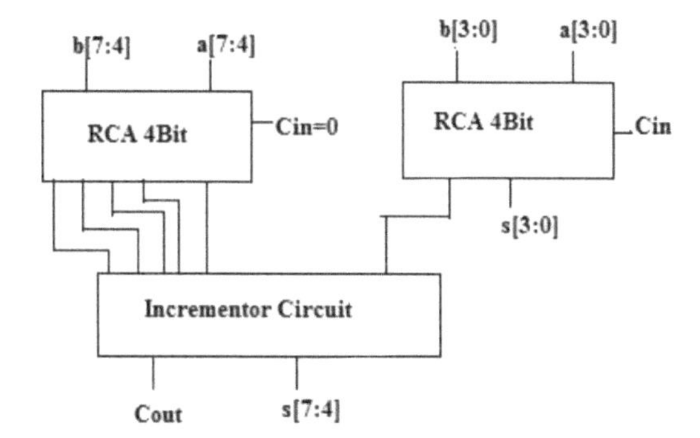

- **Ques: 261. What do you mean by carry look ahead adder?**

Solution:

Carry look ahead (CLA) design is based on the principle of looking at lower adder bits of argument and addend if higher orders carry generated. This adder reduces the carry delay by reducing the number of gates through which a carry signal must propagate.

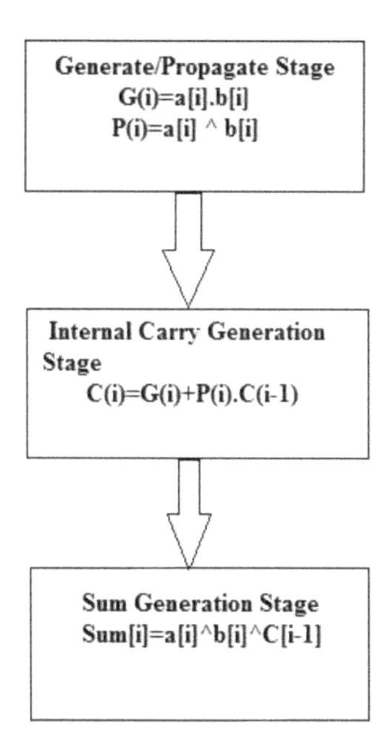

- **Ques: 262. What do you mean by carry look ahead adder?**

Solution:

In carry save adder (CSA), three bits are added parallelly at a time. In this scheme, the carry is not propagated through the stages. Instead, carry is stored in the present stage, and updated as addend value in the next stage. Hence, the delay due to the carry is reduced in this scheme.

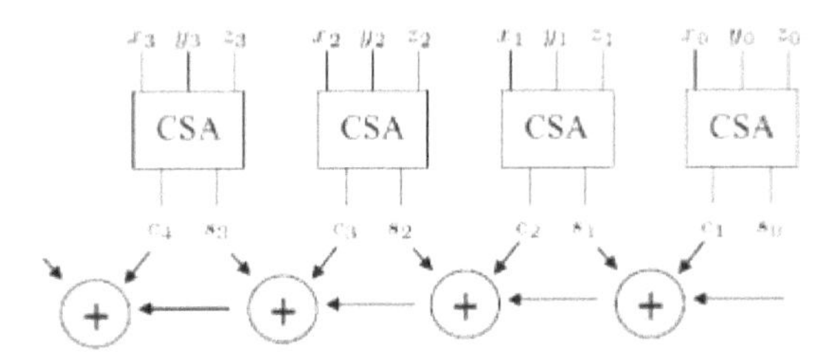

Some advanced carry look ahead architectures are the Manchester carry chain, Brent-Kung adder, and the Kogge-Stone adder.

- **Ques: 263. What do you mean by carry select adder?**

Solution:

Carry select adder (CSA) architecture consists of independent generation of sum and carry, i.e., Cin = 1 and Cin = 0 are executed parallelly. Depending upon Cin, the external multiplexers select the carry to be propagated to the next stage. Further, based on the carry input, the sum will be selected. Hence, the delay is reduced. However, the structure is increased due to the complexity of multiplexers. The architecture of CSA is illustrated in the following figure:

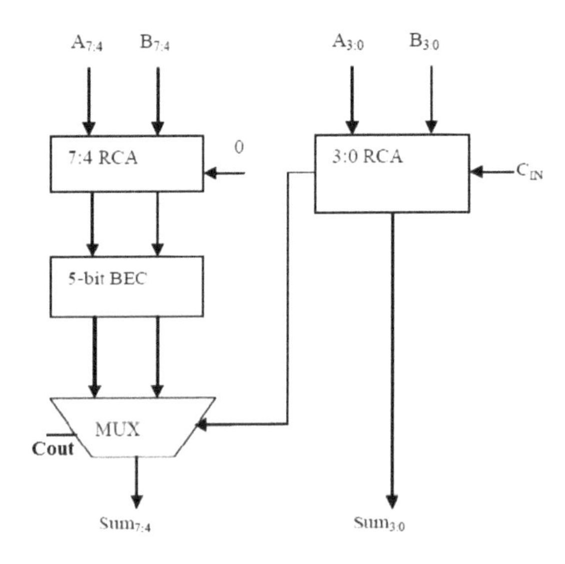

- **Ques: 264. What do you mean by carry bypass adder?**

Solution:

In carry bypass adder (CBA), RCA is used to add 4 bits at a time and the carry generated will be propagated to the next stage with the help of multiplexer using select input as bypass logic. Bypass logic is formed from the product values as it is calculated in the CLA. Depending on the carry value and bypass logic, the carry is propagated to the next stage.

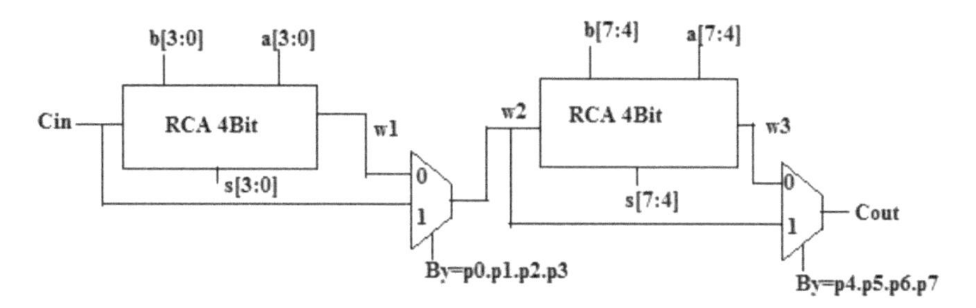

- **Ques: 285 Differentiate between latch and flip-flop.**

Solution:

S. Nos.	Latch	Flip-flop
1	It does not require any clock signal.	Flip-flop requires clock signals.
2	A latch is an asynchronous device.	A flip-flop is a synchronous device.
3	Latches are transparent devices, i.e., when they are enabled, the output changes immediately if the input is changed.	A transition from low to high or high to low clock signal will cause the flip-flop to either change its output or retain it depending on the input signal.
4	A latch is a level-sensitive device.	A flip-flop is an edge-sensitive device.
5	As there is no clock, it is simpler to design latch.	The design is relatively complex.
6	The operation of the latch is faster.	The operation of the flip-flop is relatively slower.

S. Nos.	Latch	Flip-flop
7	The power requirement is less.	The power requirement is more.
8	The latch works on the enable signal.	The flip-flop works on the clock signal.
9	It requires more calculations for timing verification.	It requires fewer calculations for timing verification.
10	The circuit analysis is comparatively difficult.	The circuit analysis is comparatively easy.
11	Time borrowing is the property of a latch by which a path ending at a latch can borrow time from the next path in the pipeline such that the overall time of the two paths remains the same. Time borrowing also is known as cycle stealing occurs at a latch.	In cycle stealing/time borrowing, flip-flops are not used.

References

1. Malvino, A. P. and Brown, J. A. (1992). *Digital Computer Electronics.* Glencoe.
2. Micheli, G. D. (1994). *Synthesis and Optimization of Digital Circuits.* McGraw-Hill Higher Education.
3. Jain, R. P. (2003). *Modern Digital Electronics.* Tata McGraw-Hill Education.
4. Keyes, R. W. (1979). The evolution of digital electronics towards VLSI. *IEEE Journal of Solid-State Circuits*, 14(2), 193–201.
5. https://www.geeksforgeeks.org/digital-electronics-logic-design-tutorials/

2

IC Fabrication Technology

In the modern age, electronics accelerates with the power of integration. Thousands and millions of transistors are combining on a single chip, to enable the user small and cheap circuit. Very large scale integration (VLSI) is a process of combining thousands of transistors into a single chip. It started in the 1970s with the development of complex semiconductor and communication technologies.

A VLSI device is commonly known as a microcontroller. Before VLSI, most integrated circuits (ICs) had limited functions. An electronic circuit usually consists of a central processor unit (CPU), read-only memory (ROM), random-access memory (RAM), and other peripherals on one board. The VLSI technology enables IC designers to compile all of these into one chip. The fact that the increasing number of components on a single chip increases with time is based on Moore's law.

2.1 Moore's Law

Since ICs came into existence in the 1960s, designers began putting dozens of components on a single chip in what was called small-scale integration or SSI. Moore made the observation and also the prediction that the number of components being placed on a chip was doubling roughly every one to two years, and would continue to do so. This is the famous Moore's law.

2.1.1 Progression of Integration Levels Through the Years

The first ICs contained only a few transistors. Early digital circuits containing transistors in tens, provided a few logic gates and early ICs had as few as two transistors. The number of transistors in an IC has increased dramatically since then.

The first ICs contained only a few transistors and so were called SSI. They used circuits containing transistors numbering in the tens. They were very crucial in the development of early computers. However, SSI was followed by the introduction of the devices which contained hundreds of transistors on each chip, and so were called "medium-scale integration (MSI)."

Further, MSI was attractive economically because it involved little additional cost for systems to be produced using smaller circuit boards, less assembly work, and several other advantages. The next development was of large-scale integration (LSI).

The development of LSI was driven by economic factors and each chip comprised tens of thousands of transistors. It was in the 1970s, when LSI started getting manufactured in huge quantities.

LSI was followed by VLSI where hundreds of thousands of transistors were used and still being developed. It was for the first time that a CPU was fabricated on a single IC to create the microprocessor. In 1986, with the introduction of the first 1 MB RAM chips, >1 million transistors were integrated. Microprocessor chips produced in 1994 contained >3 million transistors.

ULSI refers to "ultra-large scale integration" and corresponds to >1 million transistors. However, there is no qualitative leap between VLSI and ULSI; hence, normally in technical texts the "VLSI" term covers the ULSI.

2.2 Classification of Logic Families

Based on manufacturing technology, ICs are broadly categorized as follows:

(a) Bipolar families
(b) Unipolar families

Based on internal characteristics and fabrication process, they are grouped in different logic families as follows (Figure 2.1):

(a) RTL: Resistor Transistor Logic
(b) DTL: Diode Transistor Logic
(c) TTL: Transistor–Transistor Logic
(d) ECL: Emitter Coupled Logic
(e) I^2L: Integrated Injection Logic
(f) MOS: Metal–Oxide–Semiconductor

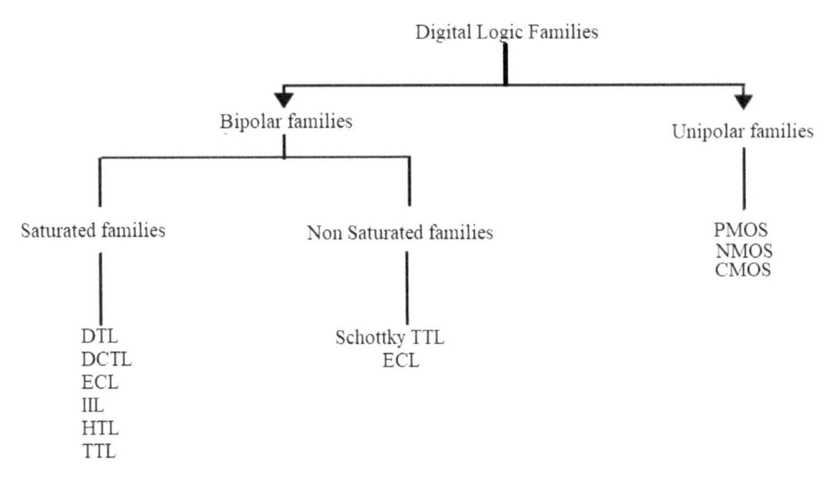

Figure 2.1 Classification of digital logic families.

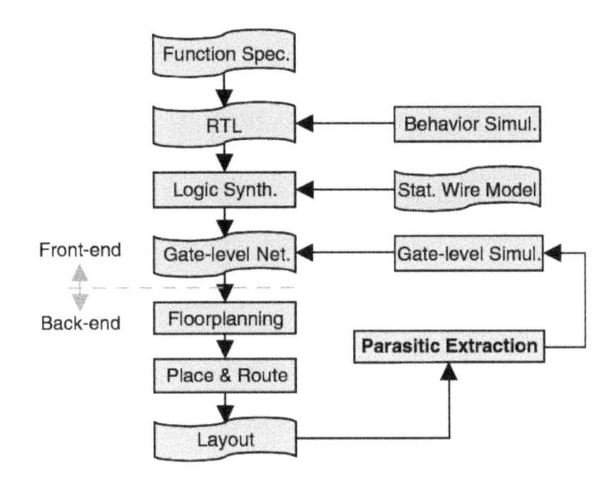

Figure 2.2 Front-end and Back-end approach.

2.3 Designing a VLSI IC

The design of a VLSI IC consists broadly of two parts, namely, front end design and back end design (Figure 2.2). The front end design includes digital design using HDLs such as Verilog, VHDL, System Verilog, and the like. It includes design verification through simulation and other verification techniques. The process includes designing, starting from gates to design for testability. The back end design comprises of CMOS library design and its characterization. It also covers physical design and fault simulation. The entire design procedure follows a step-by-step approach.

2.3.1 Front End Design Development

Problem Specification

It is a high-level representation of the system. The major parameters are performance, functionality, physical dimensions, fabrication technology, and design techniques. It has to be a trade-off between market requirements, the available technology, and the economic viability of the design. The end specifications include size, speed, power, and functionality of the VLSI system.

Architecture Definition

Basic specifications like floating-point units, which system to use, such as reduced instruction set computer (RISC) or complex instruction set computer (CISC), number of ALUs cache size, etc.

Functional Design

Defines the major functional units of the system and hence facilitates the identification of interconnect requirements between units, the physical and electrical specifications of each unit.

Logic Design

Boolean expressions, control flow, word width, register allocation, etc., are developed and the outcome is the register transfer level (RTL) description. HDLs implement the RTL description onto a system.

Circuit Design

While the logic design gives the simplified implementation of the logic, the realization of the circuit in the form of a netlist is done in this step. The netlist consists of gates, transistors, and various interconnects. This again is a software step and the outcome is checked via simulation.

Physical Design

The conversion of the netlist into its geometrical representation is done in this step and the result is called a layout. This step follows some predefined fixed rules like the lambda rules which provide the exact details of the size, ratio, and spacing between components.

Back End Design Development

Floor planning

The floor plan is the initial physical shape of the circuit. It has information about the circuit boundaries, the I/O pin location, the places where standard cells cannot be placed, and the upper metal power straps. These straps are done in upper metal to have less resistance and smaller IR drop.

Place and Route

Place and route is the back end stage that converts the gate-level netlist produced during synthesis into a physical design. Although the name denotes two phases, the place and route stage can be divided into three steps: (1) placement, (2) clock tree synthesis (CTS), and (3) routing. Placement involves placing all macros and cells into a certain and predefined space. It is done in two phases. The first one places the standard cells to optimize timing and/or congestion but not taking into account overlapping prevention. The second phase eliminates overlapping problems by placing the overlapping cells in the closest available space.

Layout and Verification

Once again, formality should be run to check the logical equivalence of the postlayout netlist with the RTL description. The huge number of transistors in a circuit can make the voltage level drop below a defined margin that ensures that the circuit works properly. IR drop analysis allows checking the power grid to ensure that it is strong enough to hold that minimum voltage level.

Parasitic extraction

Parasitic extraction has the objective to create an accurate RC model of the circuit so that future simulations and timing, power, and IR drop analysis can emulate the real circuit response. Only with this information, all the analyses and simulations can report results close to the real functioning of the circuit. This way this stage needs to precede all signoff analyses.

2.4 Types of IC Based on Manufacturing

There are two types of IC manufacturing technologies: one is the monolithic technology and the other is the hybrid technology (Figure 2.3). In the

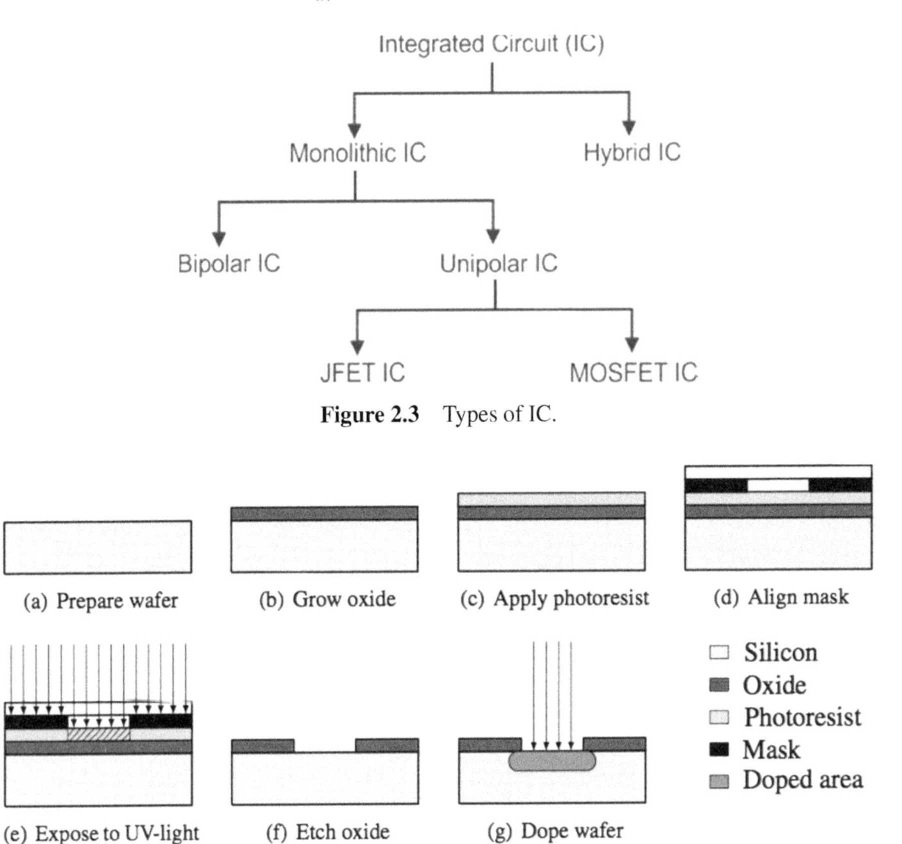

Figure 2.3 Types of IC.

Figure 2.4 Steps to fabricate an IC.

monolithic technique, all electronic components and their interconnections are manufactured together into a single chip of silicon. This technology is applied when identical ICs are to be produced on a large scale. The monolithic ICs are cheap but reliable.

In the hybrid ICs, separate components are attached to a ceramic substance and interconnected by wire or metallization pattern.

2.5 IC Fabrication Steps

Once we sort out the issues with hardware in a simulated environment, we move onto the actual hardware. Majorly talking, we have the following steps in hardware fabrication (Figure 2.4).

The monocrystal silicon wafer is polished to obtain a substrate with its surface as regular and flat as possible (a). The top of the wafer is then prepared for photolithography by covering it with an insulating layer to serve as a mask, typically an oxide (b), and a subsequent covering film of protective material which is sensitive to light, called photoresist (c). A photomask with the circuit pattern for one layer of the chip is loaded and aligned with the wafer (d). The exposure process of the wafer to intense ultraviolet (UV) light through the mask allows removing the exposed photoresist area (e). The unprotected insulating material is then stripped away using a chemical etching process and the remaining photoresist is removed by a developer solution (f). In general, there are two types of photoresist: negative and positive.

When exposed to UV light, the negative photoresist becomes polymerized and more difficult to dissolve in a developer solution than the positive resist. For the negative resist, the developer solution removes only the unexposed areas. In this way, it is possible to create a pattern of nonprotected silicon wafer areas surrounded by regions of nonconducting material. Then, the modification of the electrical properties of the exposed areas involves doping processes, such as ion implantation which is used to create sources and drains of the transistor (g).

Other conducting or insulating layers may also be added. A new layer of material is added and the entire photolithography process, which includes imaging, deposition, etching, and doping, is repeated to create many different components of the chip, layer by layer.

Wafer processing

Pure silicon is melted in a pot at 1400°C. A small seed containing the desired crystal orientation is inserted into molten silicon and slowly (1 mm/min) pulled out. The silicon crystal is manufactured as a cylindrical ingot. This cylinder is sawed into disks or wafers. Polishing and crystal orientation takes place later on.

Lithography

The process of photolithography includes masking with a photographic mask and photoetching. A photoresist film is applied to the wafer. A photoaligner aligns the wafer to a mask. Tracks are highlighted by exposing the wafer to UV light through a mask.

Etching

It removes material selectively from the surface of the wafer to create patterns. An etching mask protects some parts of the material. Additional chemicals or plasma removes the remaining photoresist. An inspection makes sure the transference of the image from the mask to the top layer of the wafer.

Ion implantation

It is a method of adding dopants. A beam of high energy dopant ions (phosphine or boron trichloride) is targeted at specific regions of a wafer. The depth of penetration into the wafer depends on the energy of the beam.

Metallization

A thin layer of aluminum is deposited over the whole wafer. Aluminum works as a good conductor and forms low resistance contacts. It can be applied and patterned with a single deposition and etching process.

Assembly and Packaging

Each of the wafers contains hundreds of chips. A diamond saw cuts the wafer into single chips separating the single chips. The chips failing electrical tests are discarded, whereas the good ones are sent for packaging. Before packaging, remaining chips are observed under a microscope. The good chips are packaged and rechecked after packaging.

2.6 Advantages and Disadvantages of IC

Although the ICs are very beneficial, they have few constraints also, which are enlisted as follows:

Advantages of IC

- Although it is quite small in size, practically around 20,000 electronic components can be incorporated in a single square inch of an IC chip.
- Many complex circuits are fabricated in a single chip and hence this simplifies the designing of a complex electronic circuit. Also, it improves performance.
- Reliability of ICs is high
- These are available at a low cost due to bulk production.

- ICs consume very minimal power.
- Higher operating speed due to the absence of parasitic capacitance effect.
- Very easily replaceable from the mother circuit.

Disadvantages of IC

- Because of its small size, the IC is unable to dissipate heat in the required rate when the current in it is increased. That is why the ICs are often damaged due to over current flowing through them.
- Inductors and transformers cannot be incorporated in the ICs.

Technical Questions with Solutions

- **Ques: 1. What do you understand by the term "IC?"**

Solution:
The expansion for IC is "integrated circuit," which can function as the microprocessor, oscillator, memory, etc. It can handle hundreds to millions of active and passive components. It is a small wafer made up of silicon. It is also called a chip or microchip.

- **Ques: 2. What is the significance of the IC number?**

Solution:
The IC number indicates the chip manufacturer as well as the specification of the IC. For example, IC number MC74HC00 indicates MC: Motorola, 74HC00: Quad-2 input NAND gate.

- **Ques: 3. How the pin number of an IC is numbered?**

Solution:
The pins are numbered anticlockwise around the IC. A dot or notch is mentioned on an IC, which is the starting point of the IC pin numbers.

- **Ques: 4. Out of the IC and discrete circuits, which are more reliable and why?**

Solution:
The ICs are more reliable than the discrete circuits because there is no need for soldering joints. The interconnections are fewer than the connections in the discrete circuits.

- **Ques: 5. Based on their application, how many types of ICs are available?**

Solution:
Based on the application area, the ICs are divided into three major classes:

1. Digital ICs
2. Analog ICs
3. Hybrid ICs

- **Ques: 6. What is a digital IC? Give any five examples of digital ICs.**

Solution:
The ICs which operate on specific logic levels (logic 0 or logic 1) rather than all levels of the signal are known as digital ICs. For example, digital logic gates, multiplexer, flip-flops, decoder, and encoder are used in the microprocessor, memory chip, computers, etc.

- **Ques: 7. What is an analog IC? Enlist any five analog ICs.**

Solution:
The ICs which operate over a continuous range of a signal are known as analog ICs. The linear ICs and radio ICs are further bifurcation of analog ICs. The five examples of analog ICs are operational amplifier, frequency mixer, linear regulator, filters, and phased locked loops.

- **Ques: 8. What do you mean by mixed-signal ICs?**

Solution:
The ICs which are fabricated by a combination of analog ICs and digital ICs on a single chip are known as mixed-signal ICs. This type of ICs is used as an analog to digital converter or digital to analog converters, timing ICs, etc.

- **Ques: 9. What is the difference between the digital IC and the analog IC?**

Solution:

1. The digital ICs can work on discrete logic levels, whereas analog ICs work on a continuous range of analog signals.
2. The digital IC design can be automated but analog IC design is very difficult, challenging, and cannot be automated.
3. The analog ICs accept an output analog data through its pins, whereas digital ICs deal with only logic data inputs and outputs.
4. Mostly, the analog IC requires external components for its functioning, whereas the digital ICs don't require external components.

- **Ques: 10. Based on the integration level, how many types of ICs are available?**

Solution:
As the number of components per chip has been increased at an accelerated pace, the levels of integration are as follows:

1. SSI (small-scale integration)
2. MSI (medium-scale integration)
3. LSI (large-scale integration)
4. VLSI (very large scale integration)
5. ULSI (ultra large scale integration)
6. GSI (Giga-scale integration)

- **Ques: 11. What is the component density in each level of integration?**

Solution:

Integration level	Number of components/gates
SSI	<12
MSI	12–99
LSI	100–9999
VLSI	10,000–99,999
ULSI	100,000–999,999
GSI	1000,000 and above

- **Ques: 12. Based on fabrication, how many types of ICs are available?**

Solution:
Based on fabrication, ICs are divided as shown hereunder:

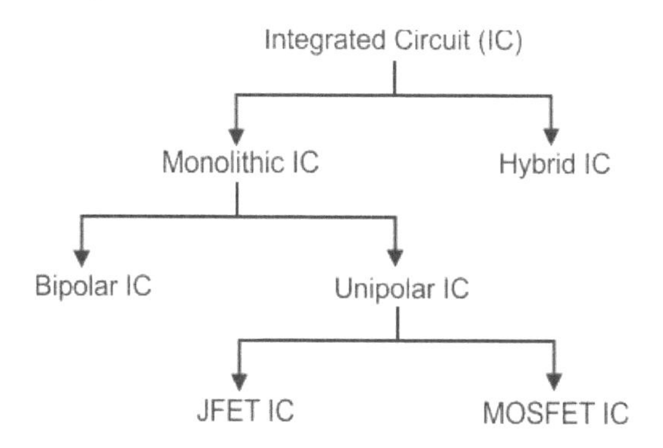

- **Ques: 13. What do you mean by monolithic and hybrid ICs?**

Solution:
In monolithic technique:

1. All electronic components and their interconnections are manufactured together into a single chip of silicon.
2. This technology is applied when identical ICs are to be produced on a large scale.
3. Monolithic ICs are cheap but reliable.

In hybrid ICs:

1. Separate components are attached to a ceramic substance and interconnected by wires or metallization pattern.
2. The hybrid ICs are costlier than the monolithic ICs.
3. The hybrid ICs are less reliable when compared with the monolithic ICs.

- **Ques: 14. Enlist the advantages and disadvantages of ICs.**

Solution:
Advantages

1. ICs are of low cost due to mass production.
2. ICs are more reliable.
3. ICs consume very little power.
4. Optimized performance due to reduced design complexity.
5. The operating speed is high.
6. Easy and economical to replace ICs from the mother circuit.
7. Very lightweight due to miniature circuit.
8. Even at high temperatures, they can withstand power fluctuations.

Disadvantages

1. If any component of the IC fails, the whole IC needs to be replaced.
2. Limited power rating
3. Delicate to handle
4. Difficult to achieve a low temperature coefficient
5. For a high value of capacitances, need exterior discrete components.

- **Ques: 15. Using planar technology, which steps are involved in IC fabrication?**

Solution:

1. Prepare silicon substrate/wafer
2. CVD (chemical vapor deposition)

3. Oxidation (silicon dioxide)
4. Photolithography
5. Diffusion
6. Ion implantation
7. Masking
8. Etching
9. Metallization
10. Packaging

- **Ques: 16. Enlist the important properties of silicon dioxide.**

Solution:

1. Silicon dioxide is an extremely hard protective coating. It is unaffected by almost all reagents except hydrochloric acid. Thus, it stands against any contamination.
2. By selective etching of silicon dioxide, diffusion of impurities through carefully defined windows in the SiO_2 can be accomplished to fabricate various components.
3. It is used as a passivating layer as the dopant impurities do not penetrate silicon dioxide. Thus, selective doping can be done using oxide as a mask.
4. It has the most stable chemical nature in comparison with germanium, which is water soluble and decomposes at high-grade temperature.

- **Ques: 17. What is the use of silicon dioxide?**

Solution:

1. In IC fabrication techniques such as etching, diffusion, ion implantation, etc.
2. In dielectrics for the electronic devices.
3. As an ultrathin layer for MOS and CMOS devices.
4. In 3D devices in MEMS technology.

- **Ques: 18. What do you mean by oxidation defects? How many types of defects may be caused during oxidation?**

Solution:
The oxidation may cause two types of faults/defects:

1. Stacking faults
2. Oxide isolation faults

Stacking faults

During oxidation, the stacking faults are induced by structural defects in the silicon lattice. The growth of stacking faults is a strong function of substrate orientation, conductivity type, and defect nuclei present. The stacking faults formation can be suppressed by the addition of hydrochloric acid.

Oxide isolation faults

The stress along the edges of an oxidized area produces severe damage in the silicon. Such defects result in increased leakage in nearby devices. High-grade temperatures will prevent stress-induced defect formation.

- **Ques: 19. What do you understand by the term "lithography?"**

Solution:
Lithography is a process by which the pattern appearing on the mask is transferred to the wafer. It involves the following two steps:

Step 1: Apply a few drops of photoresist to the surface of the wafer
Step 2: Spin the surface to get an even coating of the photoresist across the surface of the wafer. Spinner is used to coat the wafer with a uniform thick photoresist.

- **Ques: 20. Enlist the types of lithography and their use. Which technology is used for the production of masks in photolithography?**

Solution:

1. Photolithography
2. Electron beam lithography
3. X-ray beam lithography
4. Ion beam lithography

Electron beam lithography is used for the production of masks with higher resolution and shorter production time.

- **Ques: 21. Explain the process of photolithography.**

Solution:
The photolithography process mainly consists of two processes:
(a) Photographic mask
Preparation of artwork and decomposition into multilayer masks.
(b) Photoetching
Photoetching is used for the removal of silicon dioxide from specific portions so that the desired impurities can be diffused.

- **Ques: 22. What do you understand by "etching" and its types?**

Solution:
Etching is the process of material being removed from a material's surface. The two major types of etching are wet etching and dry etching. Dry etching is also known as plasma etching.

1. The etching process that involves using liquid chemicals or etchants to take off the substrate material is called wet etching. In the plasma etching process, also known as dry etching, plasmas or etchant gases are used to remove the substrate material.
2. Dry etching produces gaseous products, and these products should diffuse into the bulk gas and be expelled through the vacuum system. There are three types of dry etching (e.g., plasma etching):
   A. Chemical reactions (by using reactive plasma or gases)
   B. Physical removal (generally by momentum transfer)
   C. A combination of chemical reactions and physical removal

On the other hand, wet etching is only a chemical process.

1. Smaller line openings (1 ţm) are possible with dry etching, whereas the large line openings (>1 ţm) are possible in wet etching.

- **Ques: 23. Discuss the advantages and disadvantages of dry and wet etching.**

Solution:

Wet etching

The advantages of wet etching processes are simple equipment, high etching rate, and high selectivity.

However, there are many disadvantages.

Wet etching is generally isotropic, which results in the etchant chemicals removing substrate material under the masking material. Wet etching also requires large amounts of etchant chemicals because the substrate material has to be covered with the etchant chemical. Furthermore, the etchant chemicals have to be consistently replaced to keep the same initial etching rate. As a result, the chemical and disposal costs associated with wet etching are extremely high.

Dry etching

Some advantages of dry etching are its capability of automation and reduced material consumption. Dry etching (e.g., plasma etching) costs less to dispose of the products compared to wet etching. An example of purely chemical dry etching is plasma etching.

A disadvantage of purely chemical etching techniques, specifically plasma etching processes, is that they do not have high anisotropy because reacting species can react in any direction and can enter from beneath the masking material.

- **Ques: 24. What is an isotropic and an anisotropic etching?**

Solution:
An anisotropy is used in dry etching when the etching exclusively occurs in one direction. This property is useful when it is necessary to remove material only in the vertical direction since the material covered by the masking material would not be removed. In cases where high anisotropy is vital, dry etching techniques that use only physical removal or a combination of both physical removal and chemical reactions are used.

Isotropic etching is a wet etching process which involves undercutting. Anisotropic etching is a dry etching process that provides straight-walled patterns.

- **Ques: 25. Show the process of IC fabrication with the help of a flowchart.**

Solution:
The process of IC fabrication involves various processes, as shown in the following flowchart:

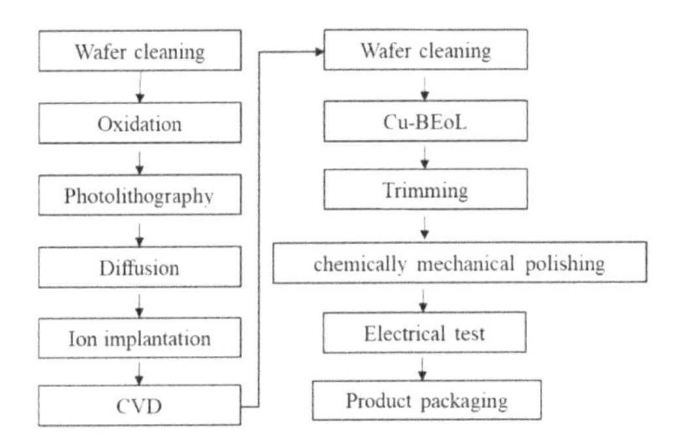

- **Ques: 26. What is reactive etching?**

Solution:
The reactive plasma is a discharge in which ionization and fragmentation of gases takes place and produce chemically active plasma species, frequently

oxidizers and reducing agents. Such plasma species are reactive both in the gas phase and with solid surfaces exposed to them.

When these interactions are used to form volatile products so that material is removed or etching of material form surfaces that are not masked to form lithographic patterns, the technique is known as reactive plasma etching.

- **Ques: 27. What do you understand by the term "diffusion?"**

Solution:
The process of introducing impurities into selected regions of a silicon wafer is called diffusion. The impurity atoms tend to move from regions of higher concentrations to lower concentrations. The rate at which various impurities diffuse into the silicon will be of the order of 1 μm/h at the temperature range of 900°C–1100°C.

- **Ques: 28. Enlist the advantages and disadvantages of ion implantation over the diffusion process.**

Solution:
Advantages of ion implantation over the diffusion process:

1. As it is performed at low temperature, therefore, previously diffused regions have a lesser tendency for lateral spreading.
2. In the diffusion process, the temperature has to be controlled over a large area inside the oven. In contrast, in the ion implantation process, accelerating potential and beam content are dielectrically controlled from outside.
3. Short process times
4. Good homogeneity and reproducibility of the profiles
5. Exact control of the number of implanted ions by measuring the current
6. Various materials can be used for masking, e.g., oxide, nitride, metals, and resist.
7. The low penetration depth of the implanted ions. This allows the modification of thin areas near the surface with high concentration gradients.

Disadvantages of ion implantation over the diffusion process:

1. The implanted ions cause damage to the substrate.
2. The change of material properties is restricted to the substrate domains close to the surface.
3. Additional effects during or after implantation, e.g., channeling or diffusion, make it difficult to achieve very shallow profiles and to theoretically predict the exact profile shapes.

- **Ques: 29. What is metallization? Which metals are commonly used for metallization?**

Solution:
The process of producing a thin metal film layer that will serve to make the interconnection of the various components on the chip is called metallization. Pure aluminum and gold are generally used in the IC fabrication process. Copper is used mostly for interconnection and metallization.

- **Ques: 30. What is the need for isolation?**

Solution:
Isolation is required to ensure that the individual devices in the chip must not interact unexpectedly. There should be no parasitic MOS conduction path present to avoid latch-up.

In dielectric isolation, a layer of solid dielectric such as silicon dioxide surrounds each component thereby producing isolation, both electrical and physical. This isolating dielectric layer is thick enough so that its associated capacitance is negligible. Also, it is possible to fabricate both PNP and NPN transistors within the same silicon substrate.

LOCOS (local oxidation of silicon) can be isolated by varying oxide thickness, which in turn limits the packaging density of the IC. Trench isolations are used nowadays to isolate the device areas.

- **Ques: 31. What are evaporation and sputtering?**

Solution:
In metallization, aluminum is deposited by sputtering or evaporation. When aluminum ions are vaporized by passing a high current through a thick aluminum wire kept inside a vacuum chamber, it is known as the evaporation technique.

When the gas plasma is generated by ionizing an inert gas by an RF electric field, the ions get focused on the aluminum container and plasma dislodges metal ions, which are further deposited on the wafer. This is known as sputtering. It is a controlled version of evaporation but having less contamination.

- **Ques: 32. Enlist the steps of silicon wafer preparation.**

Solution:

- Crystal growth and doping
- Ingot trimming and grinding
- Ingot slicing

- Wafer polishing and etching
- Wafer cleaning

- **Ques: 33. What is the use of silicon nitride in IC fabrication?**

Solution:

1. It acts as a strong barrier to most atoms.
2. Used as an overglass layer, i.e., protective coating on-chip
3. Keeps away the contaminants
4. Isolates adjacent MOSFETs
5. It enhances the storage capability of dynamic random-access memory by creating high-value capacitances.

- **Ques: 34. In layout design, which colors are commonly used for mask layers?**

Solution:
The colors that are mostly used for specific mask layers are the following:

Polysilicon: Red
Doped n^+ or p^+ (active): Green
N-well: Yellow
Metal: Blue
Contact: Black

- **Ques: 35. Is there any drawback with the aluminum metal gate?**

Solution:
As aluminum has a low melting temperature, it prohibits the use of high-temperature processes once it is deposited on the wafer. The hot electrons will get enough energy from drain to source electric field and break into silicon oxide and generate traps, which will degrade the performance of the MOS transistor.

The transistor with a silicon gate solves this problem.

- **Ques: 36. What is a silicide?**

Solution:
The highly doped polysilicon gates exhibit high resistivity. To overcome this, the poly is coated with a thin layer of refractory (high temperature) metal such as titanium (Ti), tungsten (W), or platinum (Pt). This combination is called silicide and a poly–metal mixture is treated as a single layer in the design.

- **Ques: 37. Compare aluminum and copper as an interconnect material.**

Solution:

- Copper has less resistance than aluminum. Nowadays, copper is being used in fabrication to maintain a low delay; however, capacitance increases so we have to reduce the resistance by using copper.
- Aluminum has a problem of electromigration. To overcome this problem, AlCu can be used and to prevent diffusion a barrier layer of TiN is then used.
- As copper diffuses into silicon, a barrier layer to prevent that (usually TiN) and after polishing the surface of copper, another layer is put to encapsulate the copper. This technique is called Damascene (taken from the name of the city of Damas, the capital of Syria, since they do many metal carvings there).

Aluminum has also the diffusion problem for temperatures $>450°$ (spiking), which can be reduced by replacing aluminum with AlSi.

- **Ques: 38. What is the difference between PVD and CVD?**

Solution:
The difference between CVD and PVD is the chemical reaction at the substrate surface in CVD.

PVD instead uses a vapor that can condense at the substrate's surface. In such a case, no reaction takes place. The typical PVD methods are sputtering deposition, pulsed laser deposition, and molecular beam epitaxy. The typical CVD methods are metalorganic chemical vapor deposition (MOCVD) and plasma-enhanced chemical vapor deposition (PECVD).

The CVD processes are relatively better for getting the required concentration of constituent elements. However, if the film of a single element is required, PVD techniques can safely be used and are relatively easy as compared to CVD techniques.

- **Ques: 39. What do you mean by the self-aligned gate?**

Solution:
Self-aligned signifies that there are no lithography steps involved in the specific layer creation. Since lithography is the most complex and costly process step nowadays many companies try to make as many self-aligned steps as possible. During manufacturing, the refractory gate electrode region of a MOSFET transistor is used as a mask for the doping of the source and

drain regions. This technique ensures that the gate will slightly overlap the edges of the source and drain.

- **Ques: 40. What is a passivation layer? What is its use?**

Solution:
Once all the circuit features have been formed in an IC, the chip is covered with a silicon dioxide layer known as a passivation layer. It is used to protect the chip from chemical and environmental contamination.

- **Ques: 41. Why polysilicon is used for the gate?**

Solution:
The most significant aspect of using polysilicon as the gate electrode is its ability to be used as a further mask to allow precise definition of source and drain regions. This is achieved with minimum gate to source/drain overlap, which leads to lower overlap capacitances and improved circuit performance.

1. Penetration of silicon substrate

If aluminum metal is deposited as the gate, we can't increase the temperature beyond 500 °C because aluminum will then start penetrating the silicon substrate and acts as a p-type impurity.

2. Problem with nonself-alignment

In the case of the aluminum gate, we have to first create a source and drain and then gate implant. We can't do the reverse because diffusion is a high-temperature process. And this creates parasitic overlap input capacitances Cgd and Cgs. The Cgd is more harmful because it is a feedback capacitance and hence it is reflected on the input magnified by $(k+1)$ times (recall Miller's theorem), where k is the gain.

If aluminum is used, the input capacitance increases unnecessarily which further increases the charging time of the input capacitance.

For self-alignment of gates, i.e., creation of gate followed by drain and source using ion implantation, polysilicon material is used instead of aluminum because the melting point of Al is 660 °C and doping process requires >800 °C.

- **Ques: 42. How n-well is formed?**

Solution:
N-well can be formed by using diffusion or ion implantation technique.

1. Diffusion

Place the wafer in a furnace with arsenic gas. Heat until arsenic atoms are diffused in silicon.

2. Ion implantation

Blast the wafers with arsenic ions. Ions are blocked by the silicon dioxide layer that enters it.

- **Ques: 43. What is the benefit of using n-well instead of p-well?**

Solution:
The n-well CMOS is superior to p-well because of the following reasons:

1. It lowers substrate bias effects on the transistor threshold voltage.
2. It lowers parasitic capacitances associated with the source and drain region.
3. The latch-up problems can be considerably reduced by using a low resistivity epitaxial p-type substrate.

References

1. https://electronicsforu.com/resources/learn-electronics/vlsi-developments-ic-fabrication
2. http://www.iue.tuwien.ac.at/phd/rovitto/node10.html
3. https://www.electrical4u.com/integrated-circuits-types-of-ic/
4. Gupta, P., Heng, F. L., and Lavin, M. A. (2008). *U.S. Patent No. 7,353,492*. Washington, DC: U.S. Patent and Trademark Office.
5. Bean, K. E. and Runyan, W. R. (1990). *Semiconductor Integrated Circuit Processing Technology*. Addison-Wesley.
6. Gray, P. R., Hurst, P., Meyer, R. G., and Lewis, S. (2001). *Analysis and Design of Analog Integrated Circuits*. Wiley.

3

Basic CMOS VLSI Design

One of the most popular MOSFET technologies available today is the complementary MOS or CMOS technology. This is the dominant semiconductor technology for microprocessors, microcontroller chips, memories such as RAM, ROM, EEPROM, and application-specific integrated circuits (ASICs).

The term CMOS stands for "complementary metal–oxide–semiconductor." CMOS technology is one of the most popular technologies in the chip electronics and broadly used today to form integrated circuits (ICs) in numerous and varied applications. Today's computer memories, CPUs, and cell phones make use of this technology due to several key advantages. This technology makes use of both P-channel and N-channel semiconductor devices.

3.1 Introduction

CMOS technology was invented in 1963 by Frank Wanlass while he was working at Fairchild semiconductor. CMOS is a combination of n-type and p-type MOSFET (metal–oxide–semiconductor field-effect transistor). Gordons Moore observed that the number of transistors doubles after every 18 months in an IC. This computerized electronics world demands more faster devices. This can be achievable by scaling CMOS technology from the fraction of millimeters to a few nanometers in today's technologies.

3.2 Evolution of CMOS

After bipolar junction transistor MOSFET comes with very interesting features like low power consumption, low operating voltage, higher speed, etc., which make MOSFET useful in electronics design. Two types of MOS transistor PMOS and NMOS are invented and used for designing ICs. Both types have very high static power consumption. This problem is solved if and only a logic designed in such a way that it consumes no power in a static state. After decades Frank Wanlass introduces a new logic designed using two complementary p-type and n-type MOSFETs. Two main advantages of CMOS technology are high noise immunity and very low static power consumption. The last several decades have seen the innovation of new CMOS technologies with excellent features. The trends of MOS ICs downsizing.

From the past few years, nonplanar (3D) technology by industries makes ease toward manufacturing high-speed ICs, processors, and other electronic devices. The scientist makes a very sharp reduction in the size of CMOS to 7 nm in future CMOS technologies. New 2D materials are preferred to be used for future CMOS technology. Innovations of new technologies are very important for the downsizing of CMOS ICs. As scaling down of CMOS size after every decade is difficult and we have to face some problems when downscaling of CMOS exceeds a certain limit (Figure 3.1).

As observed from Figure 3.1, the trend of downsizing of CMOS should not end. Early in 1970 CMOS is designed using planar (2D) technology. But to follow Gordons Moore's law, downsizing and new technology are very important. Now, nonplanar technology is used to design CMOS ICs.

NMOS

NMOS is built on a p-type substrate with n-type source and drain diffused on it (Figure 3.2). In NMOS, the majority carriers are electrons. When a high voltage is applied to the gate, the NMOS will conduct. Similarly, when a low voltage is applied to the gate, NMOS will not conduct. NMOS is considered to be faster than PMOS, since the carriers in NMOS, which are electrons, travel twice as fast as the holes.

(1970) (2D technology) 10µm → 8µm → 6µm → 4µm → 3µm → 2µm → 1.2µm → 0.8µm → 0.5µm → 0.35µm → 0.25µm → 180nm → 130nm → 90nm → 65nm → 45nm (2005) → 32nm (2007) → 28nm (2009) → 22nm (2012) (3D technology) → 15nm (2013) → 10nm (2015) → 7nm (2017)

Figure 3.1 Downsizing trends of MOS technology.

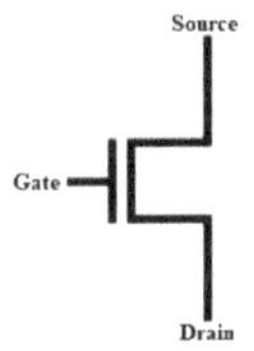

Figure 3.2 NMOS symbol.

Figure 3.3 PMOS symbol.

PMOS

P-channel MOSFET consists of p-type source and drain diffused on an n-type substrate (Figure 3.3). In the PMOS, the majority carriers are holes. When a high voltage is applied to the gate, the PMOS will not conduct. When a low voltage is applied to the gate, the PMOS will conduct. The PMOS devices are more immune to noise than the NMOS devices.

CMOS

In CMOS technology, both n-type and p-type transistors are used to design logic functions. The same signal which turns ON a transistor of one type is used to turn OFF a transistor of the other type. This characteristic allows the design of logic devices using only simple switches, without the need for a pull-up resistor.

In the CMOS logic gates a collection of n-type MOSFETs is arranged in a pull-down network between the output and the low voltage power supply rail (Vss or quite often ground) (Figure 3.4). Instead of the load resistor of the NMOS logic gates, the CMOS logic gates have a collection of p-type MOSFETs in a pull-up network between the output and the higher-voltage rail (often named Vdd).

Thus, if both a p-type and n-type transistors have their gates connected to the same input, the p-type MOSFET will be ON when the n-type MOSFET is OFF, and vice versa. The networks are arranged such that one is ON and the other OFF for any input pattern as shown in Figure 3.4.

The CMOS offers relatively high speed, low power dissipation, and high noise margins in both states, and will operate over a wide range of source and input voltages (provided the source voltage is fixed).

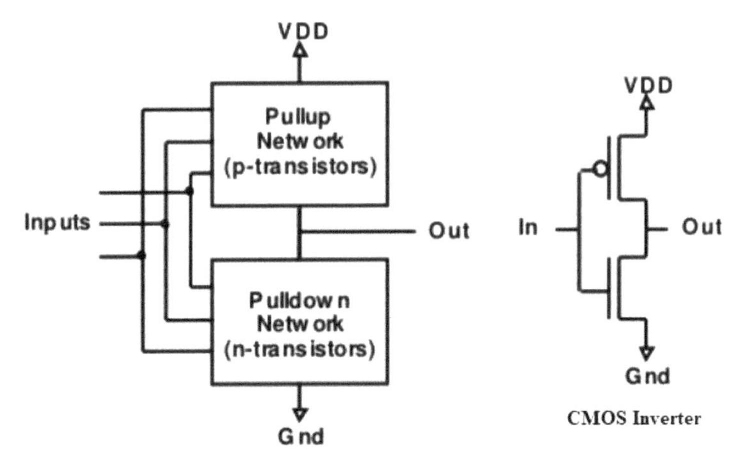

Figure 3.4 CMOS logic gate using pull-up/pull-down networks.

Technical Questions with Solutions

- **Ques: 1. What do you mean by the word "CMOS?"**

Solution:

The CMOS stands for complementary metal–oxide–semiconductor. The MOS in "CMOS" refers to the transistors in a CMOS component called MOSFETs (metal–oxide–semiconductor field-effect transistors). The "metal" part of the name is a bit misleading, as modern MOSFETs often use polysilicon instead of aluminum as the conductive material.

It is a technology to design low power ICs. The CMOS circuits are found in several types of electronic components, including microprocessors, batteries, and digital camera image sensors.

- **Ques: 2. Why CMOS is called complementary MOS?**

Solution:

The "complimentary" part of the CMOS refers to the two different types of semiconductors as each transistor contains n-type and p-type.

The n-type semiconductors have a greater concentration of electrons than holes. The p-type semiconductors have a greater concentration of holes than electrons.

These two semiconductors work together and may form logic gates based on how the circuit is designed.

- **Ques: 3. Why VLSI designers prefer MOSFET than BJT?**

Solution:

1. MOSFETs' size is smaller as compared to bipolar junction transistor (BJT).
2. The MOSFETs are easier to manufacture.
3. The MOSFETs are faster than the BJT because it has majority carriers to form current, which makes its switching faster than the BJT.
4. Large variations of Q points are not seen in the MOSFET.
5. For amplifier circuits, the MOSFETs are ideal because it has very high input impedance.
6. The MOSFETs are less noisy than the BJTs.
7. In the MOSFET, power consumption and voltage drop are lesser than the BJT.
8. Thermal runaway is lesser in the MOSFETs.
9. The MOSFET is unipolar and the BJT is bipolar.
10. The MOSFET operation is due to the input voltage and the BJT is due to the current.

- **Ques: 4. How many regions of operations are there in the MOSFET?**

Solution:

1. Cut-off region
2. Triode region/linear region/Ohmic region
3. Saturation region

- **Ques: 5. For the NMOS, what are the values of Vgs and drain current in all the regions of the MOSFET?**

Solution:

1. Cut-off region: $\mathrm{V_{gs}} < \mathrm{V_{th}}$ and $\mathrm{I_d} = 0$
2. Linear region: $\mathrm{V_{gs}} > \mathrm{V_{th}}$, $\mathrm{V_{ds}} < \mathrm{V_{gs}} - \mathrm{V_{th}}$ and $I_\mathrm{d} = \mu \cdot Cox\frac{W}{L}[(\mathrm{Vgs} - \mathrm{Vth})\mathrm{Vds} - (\frac{Vds2}{2})]$
3. Saturation region: $\mathrm{V_{gs}} > \mathrm{V_{th}}$, $\mathrm{V_{ds}} > \mathrm{V_{gs}} - \mathrm{V_{th}}$ and $\mathrm{I_d} = \frac{1}{2}\mu.Cox\frac{W}{L}(Vgs - Vth)^2$

where:

Vgs = gate to source voltage
Vds = drain to source voltage
Vth = threshold voltage
μ = electron mobility
Cox = capacitance of gate oxide
$\frac{W}{L}$ = aspect ratio of the MOSFET.

- **Ques: 6. What do you mean by threshold voltage?**

Solution:
The threshold voltage is that gate to source voltage at which the device starts to turn-on.

Vth is positive for the NMOS and negative for the PMOS.

- **Ques: 7. What is the Threshold voltage in MOSFET?**

Solution:
The threshold voltage, commonly abbreviated as Vth, of a field-effect transistor (FET), is the minimum gate-to-source voltage VGS_{th} that is needed to create a conducting path between the source and drain terminals. It is an important scaling factor to maintain power efficiency.

When referring to a junction field-effect transistor (JFET), the threshold voltage is often called "pinch-off voltage" instead. This is somewhat confusing since pinch off applied to an insulated-gate field-effect transistor (IGFET) refers to the channel pinching that leads to current saturation behavior under high source-drain bias, even though the current is never off.

- **Ques: 8. What are power dissipation and power consumption?**

Solution:
The power consumption is $Power = current \times voltage$ (P=VI)

Power dissipation is "lost" or wasted power, represented by the power converted to heat that is no longer useful in that system. The dissipated power is the difference between the input and the output power, as the rest is wasted or dissipated. So, power dissipation will be approx. 10% of the power consumption on full load.

- **Ques: 9. What is the pinch-off voltage in the FET?**

Solution:
As the gate to source voltage is made more negative, the width of the channel decreases until no more current flows between the drain and the source, then the FET is said to be "pinched-off." The voltage at which the channel closes is called the pinch-off voltage.

So, the pinch-off voltage is the voltage of drain current after which the drain current is kept constant. Vgs is also known as the cut-off voltage, the minimum current required to turn the JFET off.

- **Ques: 10. How the threshold voltage in the CMOS circuits can be reduced for low power applications?**

Solution:
The threshold voltage can be reduced by:

1. Reducing the oxide thickness.
2. Reducing the channel length (short channel effect) threshold voltage is reduced
3. Increasing the body and drain voltage.
4. By controlling ion implantation of the suitable doping atoms for both n- and p-channel transistors.

- **Ques: 11. In the CMOS inverter, what will happen if the NMOS is connected to the Vdd and the PMOS is connected to the Vss?**

Solution:
An NMOS can conduct a strong 0 but a weak 1 while the PMOS conducts a strong 1 but a weak 0. When the PMOS and NMOS are interchanged in a CMOS inverter, it results in a degraded buffer with weak output states.

Let's understand the scenario with more deliberation:

PMOS passes good "1"-Vdd
NMOS passes good "0"-and

NMOS – works when the input to the gate is high.
PMOS – works when the input to the gate is low.

The threshold is +ve for the NMOS and –ve for the PMOS.

The source of the PMOS is connected to the VDD and the NMOS to the VSS.

In the figure given below

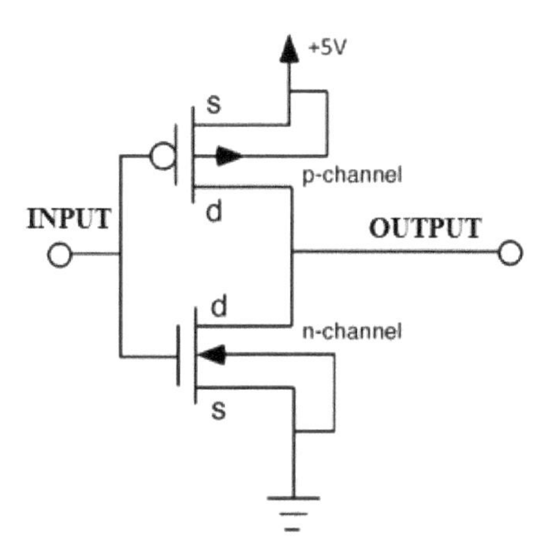

We know current will always conduct when there will be a channel, i.e., Vgs > Vt (threshold voltage)

Suppose Vt = 1 volt

Case 1: For PMOS

For Pmos: (Vt = −1 volt for Pmos Vt is negative) When Vs is connected to Vdd:

Vg = 0; Vs = 5; Vgs = Vg − Vs = −5 volt since |Vgs|> |Vt|, it conducts normally

i.e., Vout = Vdd (5 volt)-no drop, full efficiency

When Pmos connections are interchanged:

Vs = Gnd = 0
Vg = o (eq2)
Vgs = 0 volt
|Vgs|< |Vt|
No conduction it cannot 0 volt
That is why Pmos is called good 1 (Vdd connected to vs)

Case 2: For NMOS

Vt = 1 volt, for NMOS Vt is positive

When in normal operation suppose
Vg = 5 v and
Vs = 0 v and

For the above-mentioned case Vgs = Vg − Vs = 5 v, which is greater than the threshold voltage and logic 0 will easily pass.

Now if we want to pass logic 1, we have to make Vs = Vdd (suppose 5 v).

In this case, Vgs = 0 V which is less than the threshold voltage, thus channel is not formed. And a bad logic 1 is passed. rather than Vout = Vdd, there will be a loss equivalent to Vt.

Vout = Vdd − Vt (full efficiency is not achieved)

So, NMOS is a good 0″(gnd) and bad 1(vdd).

Degraded operation because see we are not getting output as Vdd, we are getting (Vdd–Vt) a loss of efficiency.

We get large swing if NMOS used as pull down and PMOS as pull up.

- **Ques: 12. Why PMOS is connected to Vdd and NMOS is connected to ground?**

Solution:

1. To prevent latch-up in the CMOS, the body-source and body-drain diodes should not be forward-biased, i.e., body terminal should be at the same or lesser voltage than source terminal (for an NMOS) and for the PMOS, it should be at a higher voltage than the source. This condition will be satisfied if we connect all the NMOS to their respective sources, i.e., ground. Similarly, the body of all the PMOS transistors is connected to a common terminal, Vdd.
2. The PMOS passes strong 1 but weak 0 and the NMOS passes strong 0 but weak 1. So, if you connect the NMOS to 1, it passes only VDD – Vtn, So, its working will be not as per expectation. similarly, for the PMOS vice versa.
3. Consider the NMOS structure, p-substrate, and two n+ diffusions are there, from downsides, it looks like a p–n junction. when you connect it to 1, it becomes forward bias and the NMOS is not in the control of gate voltage, So, the NMOS is always tied to ground 0 and the PMOS for VCC or 1.

- **Ques: 13. Out of NMOS and PMOS, which is used more for fabrication? and why?**

Solution:
Out of NMOS and PMOS, the designers prefer NMOS more. The reasons for preferring NMOS are the following:

1. NMOS is faster than PMOS as mobility of carriers (electrons) in NMOS greater than that of holes in PMOS.
2. As conductivity is proportional to mobility, NMOS offers fewer RDS (ON) and gate capacitance than PMOS for the same die area.
3. As the mobility of the carriers in an NMOS is $\sim$2–3 times higher than that of a PMOS, for the same Rds (ON) value, the PMOS must be 2–3 times the size of the NMOS.
4. In high-frequency applications, PMOS results in higher switching losses due to their higher gate charge requirement than NMOS for the same current rating.
5. PMOS takes more space and it is harder to fabricate than NMOS.

- **Ques: 14. What do you mean by Vdd and Vss?**

Solution:
Vdd: voltage (at) drain
Vss: voltage (at) source

The bibliographic reference suggests that when the same subscript is repeated, it refers to the power supply. For example, Vdd, here it refers to positive supply and Vss as a negative power supply.

- **Ques: 15. What do you understand by body effect?**

Solution:
Body effect refers to the change in the transistor threshold voltage (Vth) resulting from a voltage difference between the transistor source and body. Because the voltage difference between the source and body affects the Vth, the body can be thought of as a second gate that helps determine how the transistor turns on and off.

- **Ques: 16. What do you mean by body bias?**

Solution:
Body bias involves connecting the transistor bodies to a bias network in the circuit layout rather than to power or ground. The body bias can be supplied from an external (off-chip) source or an internal (on-chip) source.

In MOSFET second-order effect analysis, it is assumed that the source and the bulk of the transistor are tied together. The voltage difference between the source and bulk (substrate) (Vsb) affects Vth of the transistor as per Shichman-Hodges equation:

$$V_{TN} = V_{TO} + \gamma(\sqrt{|V_{SB} + 1\phi_F|} - \sqrt{|2\phi_F|})$$

So, body bias is a voltage at which the body terminal (4th terminal) is connected. When the body of the transistor is not biased at the same level as the source, it is known as body effect.

Body bias is a potential difference between substrate and source.

Body effect is increased in Vt because of body bias.

- **Ques: 17. Why is the number of gate inputs to CMOS is usually limited to four?**

Solution:
The more transistors you have, the more capacitance they introduce, and hence the long delay. To offset this, designers will have to increase the transistor sizes.

1. The number of inputs is equal to several transistors in the stack. So, it is directly proportional to the height of slack. With the increase in the

number of transistors, the speed will be slower, so it is limited to four (maximum).
2. The maximum number of gate inputs is limited due to electrical constraints.
3. The greater the number of inputs, the higher the resistance to draw the NAND output from high to low.
4. Size and delay factor.

- **Ques: 18. What are the three main aspects of VLSI optimization?**

Solution:

1. Power
2. Area
3. Speed

- **Ques: 19. What is the need for low power devices?**

Solution:
With the reduction in power:

1. Noise immunity increases
2. Increases battery lifetime
3. Cooling is reduced
4. Packaging cost is reduced

- **Ques: 20. Enlist a few low power design circuit designs.**

Solution:

1. Voltage scaling
2. Transistor resizing
3. Pipelining and parallelism
4. Power management modes

- **Ques: 21. What are the sources of power dissipation in CMOS?**

Solution:

1. Dynamic power consumption
   Charging and discharging capacitors
2. Short circuit currents
   Short circuit path between supply rails during switching
3. Leakage
   Leaking diodes and transistors

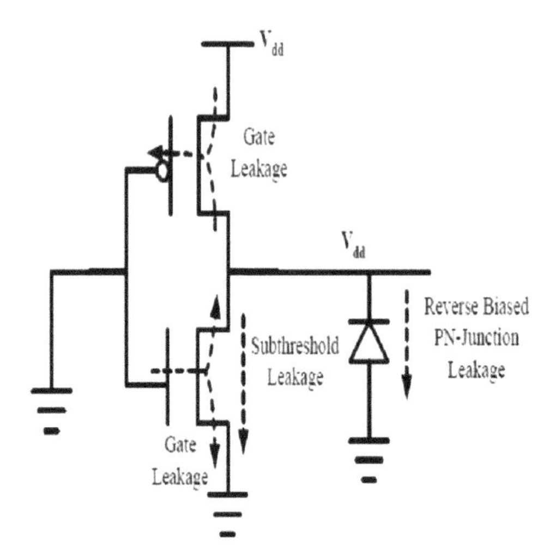

The leakage current of a transistor is mainly the result of reverse-biased P–N junction leakage, subthreshold leakage, and gate leakage.

- **Ques: 22. What is dynamic power dissipation in CMOS?**

Solution:

The dynamic power dissipation is the dynamic power dissipation, also called the switching power. This is the dominant source of power consumption in CMOS system-on-chip (SoC), accounting for roughly 75% of the total.

$$Dynamic\ Power = \alpha \cdot Cl \cdot f \cdot Vdd^2$$

where:

$\alpha =$ switching activity factor

$Cl =$ overall capacitance to be charged and discharged in a reference clock cycle. Technology scaling has resulted in smaller transistors and hence smaller transistor capacitances, but interconnect capacitance has not scaled much with the process and has become the dominant component of capacitance.

$Vdd =$ supply voltage. Though voltage scaling has the biggest impact on power dissipation (nearly quadratic savings in power), this generally comes at the expense of an increase in delay.

$f =$ Switching frequency of a global clock for a globally synchronous design, local clock for a locally synchronous design or the input arrival rate in case of a purely static system.

- **Ques: 23. Enlist the techniques for leakage power reduction.**

Solution:

1. Optimal standby input vectors
2. Dual-*Vth* assignment
3. Multithreshold-voltage CMOS
4. Super Cut-off CMOS technique (SCCMOS)
5. SLEEP transistor technique
6. SLEEPY stack technique
7. SLEEPY KEEPER technique
8. Transistor stacking
9. Forced stack
10. Variable threshold CMOS (VTCMOS)
11. Multipurpose technique
12. ZIGZAG technique
13. LEAKAGE FEEDBACK technique
14. GALEOR technique

- **Ques: 24. What do you mean by optimal standby input vectors?**

Solution:
Subthreshold leakage current depends on the vectors applied to the gate inputs because different vectors cause different transistors to be turned off. When a circuit is in the standby mode, one could carefully choose an input vector and let the total leakage in the whole circuit to be minimized model leakage current using linearized pseudo-Boolean functions.

- **Ques: 25. What is the Dual-Vth Assignment technique?**

Solution:
The dual-Vth assignment is an efficient technique for leakage reduction. In this method, each cell in the standard cell library has two versions, low Vth and high Vth. Gates with low Vth are fast but have high subthreshold leakage, whereas gates with high Vth are slower but have much reduced subthreshold leakage. Traditional deterministic approaches for dual-threshold assignment utilize the timing slack of noncritical paths to assign high Vth to some or all gates on those noncritical paths to minimize the leakage power.

- **Ques: 26. What is Multithreshold-voltage CMOS?**

Solution:
A multithreshold-voltage CMOS (MTCMOS) circuit is implemented by inserting high Vth transistors between the power supply voltage and the

original transistors of the circuit. The original transistors are assigned low Vth to enhance the performance while high-Vth transistors are used as sleep controllers. Inactive mode, SL is set low and sleep control high-Vth transistors (MP and MN) are turned on. Their on-resistance is so small that VSSV and VDDV can be treated as almost being equal to the real power supply. In the standby mode, SL is set high, MN and MP are turned off and the leakage current is low. The large leakage current in the low Vth transistors is suppressed by the small leakage in the high-Vth transistors. By utilizing the sleep control high-Vth transistors, the requirements for high performance in active mode and low static power consumption in standby mode can both be satisfied.

- **Ques: 27. What is super cut-off CMOS technique (SCCMOS)?**

Solution:
This technique is very much similar to the MTCMOS but instead of a high Vth sleep transistor, a nominal Vth sleep transistor is employed to reduce the additional delay caused due to the presence of increased threshold value in sleep transistor.

- **Ques: 28. What is the SLEEP transistor technique?**

Solution:
This is a state-destructive technique that cuts off either pull-up or pull-down or both the networks from supply voltage or ground or both using sleep transistors. This technique is MTCMOS, which adds high-Vth sleep transistors between pull-up networks and Vdd and pulldown networks and gnd while for fast switching speeds, low-Vth transistors are used in logic circuits. Isolating the logic networks, this technique dramatically reduces leakage power during sleep mode. However, the area and delay are increased due to additional sleep transistors. During the sleep mode, the state will be lost as the pull-up and pull-down networks will have floating values. These values impact the wakeup time and energy significantly due to the requirement to recharge transistors which lost state during sleep.

- **Ques: 29. What is the SLEEPY STACK technique?**

Solution:
This technique combines the structure of the forced stack technique and the sleep transistor technique. In the sleepy stack technique, one sleep transistor and two half-sized transistors replace each existing transistor. Although using $W_0/2$ for the width of the sleep transistor, changing the sleep transistor width

may provide additional trade-offs between delay, power, and area. It also requires additional control and monitory circuit for the sleep transistors.

- **Ques: 30. What is the SLEEPY KEEPER technique?**

Solution:
This technique consists of sleep transistors connected to the circuit with NMOS connected to Vdd and PMOS to GND. This creates virtual power and ground rails in the circuit, which affects the switching speed when the circuit is active. The identification of the idle regions of the circuit and the generation of the sleep signal need additional hardware capable of predicting the circuit states accurately, increasing the area requirement of the circuit. This additional circuit consumes power throughout the circuit operation to continuously monitor the circuit state and control the sleep transistors even though the circuit is in an idle state.

- **Ques: 31. What is transistor stacking?**

Solution:
The two serially connected devices in the off state have significantly lower leakage current than a single off device. This is called the stacking effect. With transistor stacking by replacing one single off transistor with a stack of serially connected off transistors, leakage can be significantly reduced. The disadvantages of this technique are also obvious. Such a stack of transistors causes either performance degradation or more dynamic power consumption.

- **Ques: 32. What is forced stack?**

Solution:
In this technique, every transistor in the network is duplicated with both the transistors bearing half the original transistor width. Duplicated transistors cause a slight reverse bias between the gate and source when both transistors are turned off. Because the subthreshold current is exponentially dependent on gate bias, it obtains substantial current reduction. It overcomes the limitation with sleep technique by retaining state but it takes more wakeup time.

- **Ques: 33. What is variable threshold CMOS (VTCMOS)?**

Solution:
Variable threshold CMOS (VTCMOS) is a circuit design technique that has been developed to reduce standby leakage currents in low VDD and low VT applications. Rather than employing multiple threshold voltage process options, a VTCMOS circuit inherently uses low threshold voltage transistors,

and the substrate bias voltages of the NMOS and PMOS transistors are generated by the variable substrate bias control circuit.

- **Ques: 34. What is the difference between BJT and MOSFET?**

Solution:
The few differences between FET and BJT are enlisted as follows:

S.No.	Field Effect Transistor	Bipolar Junction Transistor
1	The voltage gain is very low	The voltage gain is high
2	The current gain is high	The current gain is low
3	Very high input impedance, because no current flows through the gate, it is in the range of 10^6 ohms. So, it is an excellent choice for amplifier inputs.	Low-medium input impedance (1k–3k ohms)
4	Very high output impedance	Low output impedance
5	The switching time is fast	It has a medium switching time
6	It is a voltage-controlled device.	It is current-controlled device.
7	It is expensive than bipolar	It is cheaper
8	FET is more stable to temperature, so used in high-temperature applications	BJT is temperature-sensitive.
9	FET is smaller in size	BJT is bigger in size
10	FET is having a positive temperature coefficient.	BJT is having a negative temperature coefficient, so it is liable for thermal runaway.
11	Once the gate terminal of FET is charged, no more current is required to keep the transistor ON.	It requires a small amount of current to make it ON, the heat dissipated limits the number of transistors on the chip.
12	Noise level is low	Noise level is high
13	FETs are preferred in wide line or load variations and have a low power consumption.	BJTs are used where high gain and fast response is required.

- **Ques: 35. Differentiate between BIOS and CMOS.**

Solution:
The few differences between BIOS and CMOS are enlisted as follows:

S. No.	BIOS	CMOS
1	A nonvolatile firmware that initializes hardware during the booting process and provides runtime services for operating systems and programs.	It is a special memory chip in the motherboard that stores and holds BIOS configuration settings.
2	BIOS stands for basic input output system	CMOS stands for complimentary metal–oxide–semiconductor
3	It is nonvolatile	It is volatile
4	It initializes hardware while booting up the computer and provides runtime services for operating system and programs.	It stores all BIOS settings.

- **Ques: 36. What is the multipurpose/hybrid technique?**

Solution:
This technique has a combined structure of the following techniques:

- MTCMOS technique
- SCCMOS technique
- Forced stack technique

Unlike the MTCMOS and SCCMOS technique, this technique can reduce power consumption during active mode and retain the exact logic state and unlike the forced stack technique, this technique can save power consumption during standby mode.

- **Ques: 37. What is the ZIGZAG technique?**

Solution:
The wake-up cost can be reduced in the zigzag technique but still state losing is a limitation. Thus, any particular state which is needed upon wakeup must be regenerated somehow. For this, the technique may need extra circuitry to generate a specific input vector.

- **Ques: 38. What is the leakage feedback technique?**

Solution:
This technique is based on the sleep approach. To maintain logic during sleep mode, the leakage feedback technique uses two additional transistors and the two transistors are driven by the output of an inverter which is driven by the output of the circuit implemented utilizing leakage feedback. Performance degradation and increase in the area are the limitations along with the limitation of sleep technique.

- **Ques: 39. What is the GALEOR technique?**

Solution:
In the GALEOR technique, two gated leakage transistors are inserted between pull-up and pull-down networks of the CMOS circuit. A gated leakage NMOS transistor is placed between output and pull-up circuit and a gated leakage PMOS transistor is placed between output and pull-down circuitry. The gates of these additional transistors are controlled by the drain voltages.

Gated leakage transistors cause an increase in resistance of the path from Vdd to ground since one of the leakage transistors is always near its cut-off region, thereby decreasing leakage current. GALEOR technique reduces the leakage current to some extent but it suffers from a significant problem of low voltage swing. In this technique, the logic low level is very much higher than 0 volts and the logic high level is much lower than Vdd, which makes voltage swing to be lowered. This reduced voltage swing increases the propagation delay through the circuit.

- **Ques: 40. What do you understand by noise margin?**

Solution:
Noise margin is a parameter closely related to the input–output voltage characteristics. This parameter allows us to determine the allowable noise voltage on the input of a gate so that the output will not be affected. The specification most commonly used to specify noise margin (or noise immunity) is in terms of two parameters—the LOW noise margin, NML and the HIGH noised margin, NMH. Consider the figure that follows:

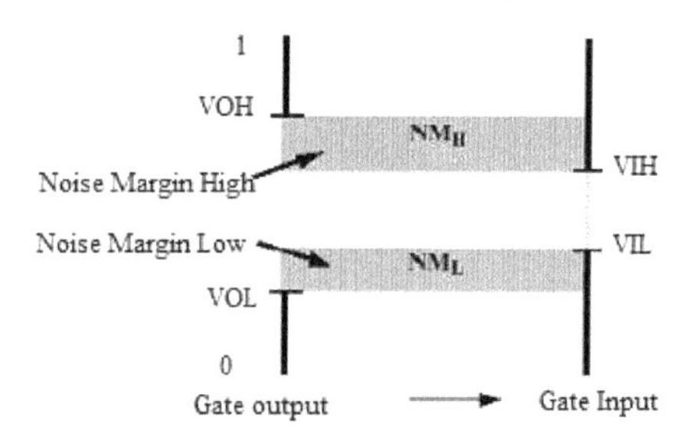

NML is defined as the difference in magnitude between the maximum LOW output voltage of the driving gate and the maximum input LOW voltage recognized by the driven gate. Thus,

$$NML = |VILMax–VOLMax|$$

NMH is the difference in magnitude between the minimum HIGH output voltage of the driving gate and minimum input HIGH voltage recognized by the receiving gate.

$$NMH = |VOHMin–VIHMin|$$

- **Ques: 41. How to reduce heat in CMOS circuits?**

Solution:
As we know the dynamic power in CMOS is

$$Dynamic\ Power = \alpha \cdot Cl \cdot f \cdot Vdd^2$$

The heat can be reduced by

1. Reducing clock frequency
2. Reducing load capacitance
3. Reducing rail voltage, Vdd
4. Reducing the switching activity parameter
5. Using pipelining to operate internal logic at a lower clock than I/O frequency

- **Ques: 42. What do you understand by "pseudo-NMOS?"**

Solution:
The inverter that uses a p-device pull-up or load that has its gate permanently grounded. An n-device pull-down or driver is driven with the input signal. This is roughly equivalent to the use of a depletion load in NMOS technology and is thus called "pseudo-NMOS."

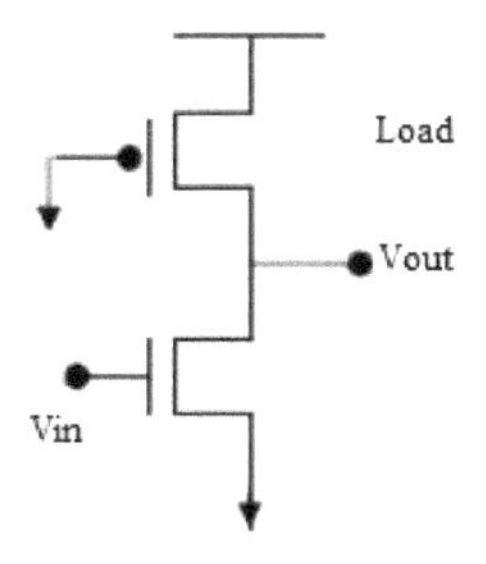

The circuit is used in a variety of CMOS logic circuits. In this, PMOS for most of the time will be linear region. So, the resistance is low and hence RC time constant is low. When the driver is turned on, a constant DC flows in the circuit.

- **Ques: 43. What are the advantages and disadvantages of pseudo-NMOS logic?**

Solution:
It is also known as ratioed logic. The advantages and disadvantages are enlisted as follows:
Advantages

1. Only n + 1 transistor is required for the n-input gate. So, the area cost is low.
2. The gate-load capacitance is low.
3. Higher speed
4. Less power

Disadvantages

1. Nonzero static power dissipation
2. Cannot be operated at slow speed
3. It requires a clock
4. Circuits are more sensitive to timing error and noise
5. It is affected by charge sharing
6. Design is more difficult

- **Ques: 44. Realize exclusive OR gate using pseudo-NMOS logic?**

Solution:

$$Y = X_1 \oplus X_2 = X_1\overline{X_2} + \overline{X_1}X_2 = \overline{X_1X_2 + \overline{X_1}\,\overline{X_2}} = \overline{X_1X_2 + \overline{X_1 + X_2}}$$

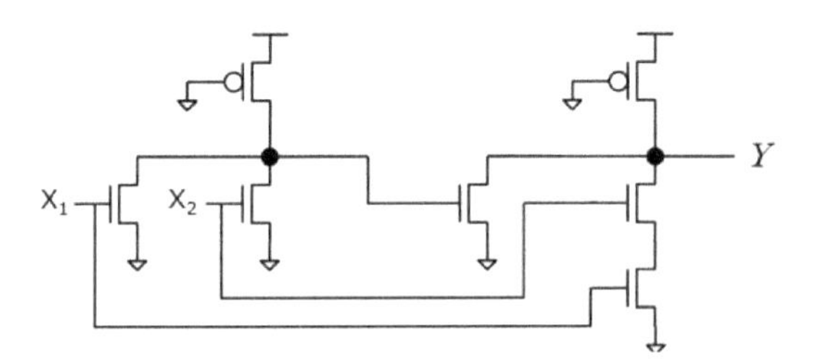

- **Ques:45. How static power in CMOS is calculated?**

Solution:
A pseudo-NMOS logic gate having a "1" output has no static (DC) power dissipation.

However, a pseudo-NMOS gate having a "0" output has a static power dissipation. The static power dissipation is equal to the current of the PMOS load transistor multiplied by the power supply voltage.

Larger the PMOS, the larger the power dissipation. The static power dissipation in CMOS is given as

$$P_{dc} = \frac{\mu_p C_{ox}}{2} \left(\frac{W}{L}\right)_P (V_{gs} - V_{tp})^2 V_{dd}$$

- **Ques: 46. How static power can be reduced using dynamic logic in CMOS?**

Solution:
The dynamic logic is used to reduce static power. It has two modes as listed hereunder:

1. Using dynamic precharging
   Normally, during the time the output is being precharged, the NMOS network should not be conducting.

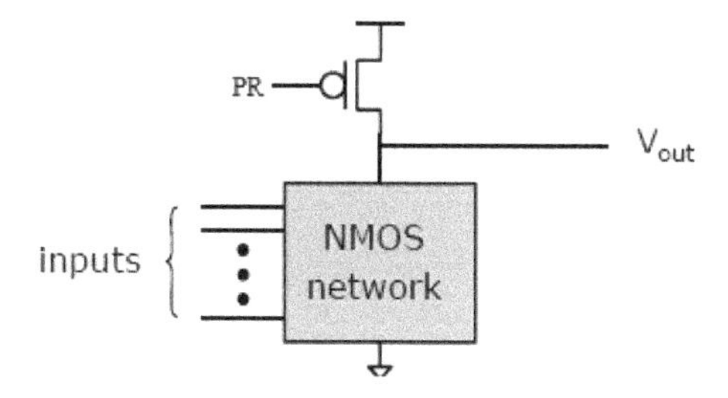

   But this is usually not possible.
2. Using precharge and evaluation

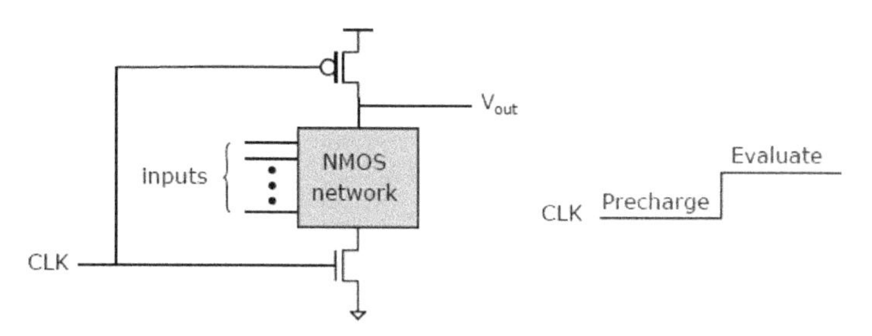

It has a two-phase operation: precharge and evaluates. It can reduce the static power dissipation to a great extent.

In single-phase dynamic CMOS logic:
Precharge phase, CLK = 0
Evaluate phase, CLK = 1

The problem with this model is that input can be changed only during the precharging phase, and if they change during the evaluation, charge redistribution can corrupt the output voltage.

- **Ques: 47. Design the following logic using pseudo-NMOS.**
    - **Y = ABC**
    - **Z = (A + B).C**

Solution:
Using pseudo-NMOS logic, the designs are as per the following figures:

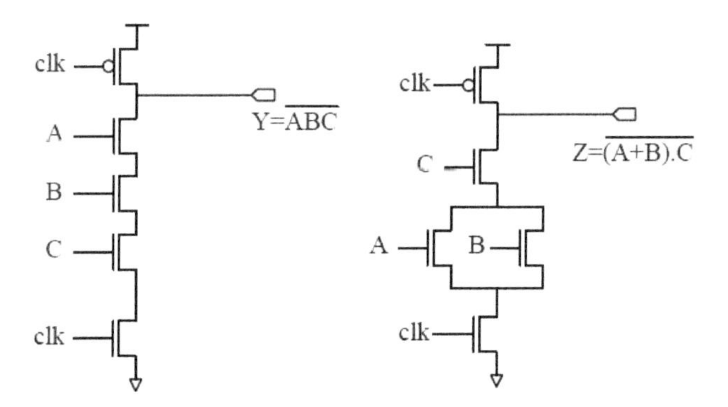

(a) Y = ABC (b) Z = $(A + B).C$

- **Ques: 48. What are the features of static and dynamic CMOS logic?**

Solution:
In static circuits, at every point in time (except when switching) the output is connected to either GND or VDD via a low resistance path. Further, fan-in of n requires 2n devices. Dynamic circuits rely on the temporary storage of signal values on the capacitance of high impedance nodes. requires only n + 2 transistors take a sequence of precharge and conditional evaluation phases to realize logic functions.

1. Charge sharing
2. Cascading problem of single-phase dynamic logic

In dynamic CMOS output is taken across a capacitor. The output is not always connected to supply or GND. Charge loss on the capacitor is the main issue in dynamic CMOS. It needs a timely refresh to store charge on the capacitor.

- **Ques: 49. What is charge sharing in dynamic CMOS logic?**

Solution:
The charge-sharing problem occurs when the charge which is stored at the output node in the precharge phase is shared among the junction capacitance of transistors in the evaluation phase. Charge sharing may degrade the output voltage level or even cause erroneous output value.

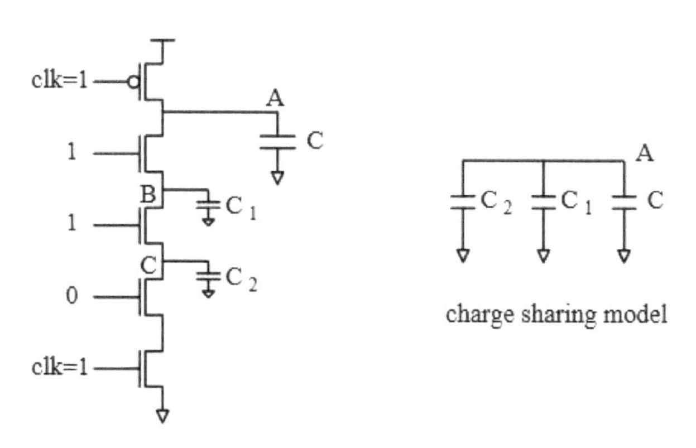

$$CV_{DD} = (C + C_1 + C_2)V_A$$

$$V_A = \frac{C}{C + C_1 + C_2} V_{DD}$$

If $C_1 = C_2 = 0.5$ V, then the output voltage is $\frac{VDD}{2}$

In the dynamic CMOS circuit technique, the clock pulse is given between a PMOS and an NMOS and the NMOS logic is associated with them. The circuit operation is based on first precharging the output node capacitance and subsequently estimating the output level according to the applied inputs. Together, these operations are scheduled by a single clock signal which drives one NMOS and one PMOS transistors in each dynamic stage. When the clock signal $= 1$
then precharge transistor p1 turns OFF and n1 turns ON. If the input signal forms a conducting path between the output node and ground then output capacitance will discharge to 0 V. When the clock signal $= 0$
then the PMOS transistor p1 is conducting and the complementary NMOS

transistor n1 is OFF. The output capacitance of the circuit is charged up through the conducting PMOS transistor to a logic high level of VDD.

- **Ques: 50. How the charge-sharing problem can be reduced?**

Solution:
To eliminate charge-sharing problem is just to add a weak PMOS pull-up device (with a small W/L ratio) to the dynamic CMOS stage output, which forces a high output level except there is a strong pull-down path amongst the output and the ground. It can be observed that the weak PMOS transistor will be turned on only when the precharge node voltage is retained high. Otherwise, it will be turned off as the output voltage becomes high.

- **Ques: 51. Show that single-phase dynamic logic cannot be cascaded.**

Solution:
When a single-phase clock is used to cascade domino logic, it produces an error state, as shown in the following figure:

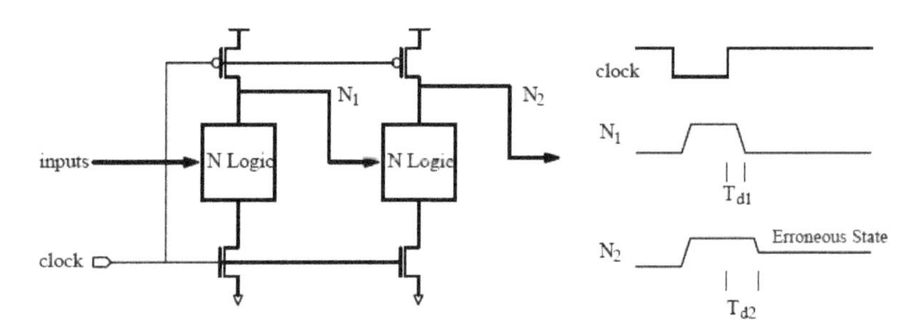

- **Ques: 52. What do you mean by domino logic? What are its properties?**

Solution:
Domino logic is a CMOS-based evolution of the dynamic logic techniques. It allows a rail-to-rail logic swing. It was developed to speed up circuits. The problem of single-phase cascading is resolved using domino logic. The basic structure of domino logic is:

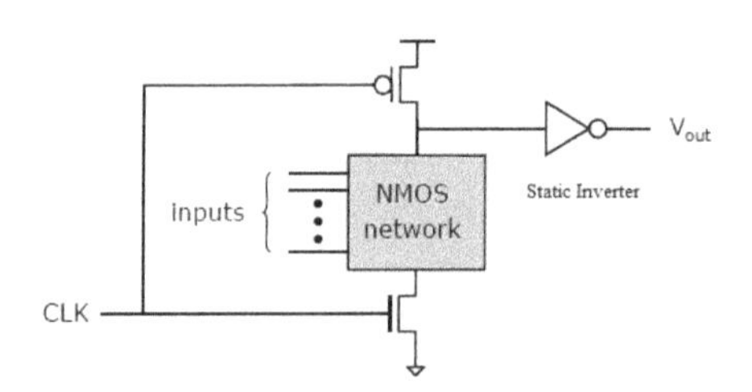

Domino CMOS has the following properties:

- Each gate requires n + 4 transistors
- Logic evaluation propagates as falling dominoes; hence, minimum evaluation period is determined by the logic depth.
- The nodes must be precharged during the precharge period. Total precharge time depends on the size of PMOS.
- Inputs must be stable (only one rising transition) during the evaluation period.
- Gates are ratio-less and are noninverting.

Its advantage is high speed and low device count. But it has the drawback of noise immunity due to leakage current and charge sharing. So, a PMOS keeper must be used to compensate for this leakage. However, the use of a keeper in the conventional domino circuit degrades the speed of the circuit or results in an erroneous output due to the contention current.

- **Ques: 53. What is the advantage of introducing a static inverter in domino logic?**

Solution:
The introduction of the static inverter has the additional advantage of the output having a low impedance output, which increases noise immunity and drives the fan-out of the gate. The buffer furthermore reduces the capacitance of the dynamic output node by separating internal and load capacitance. The buffer itself can be optimized to drive the fan-out in an optimal way for high speed. The inverter will introduce another problem that this type of logic family is noninverting.

- **Ques: 54. What is the "charge keeper circuit" in domino logic?**

Solution:
The domino cascade must have an evaluation interval that is long enough to allow every stage time to discharge. This means that charge sharing and charge leakage processes that reduce the internal voltage maybe limiting factors.

This can be reduced using the following two versions:

1. Static version
2. Latched version

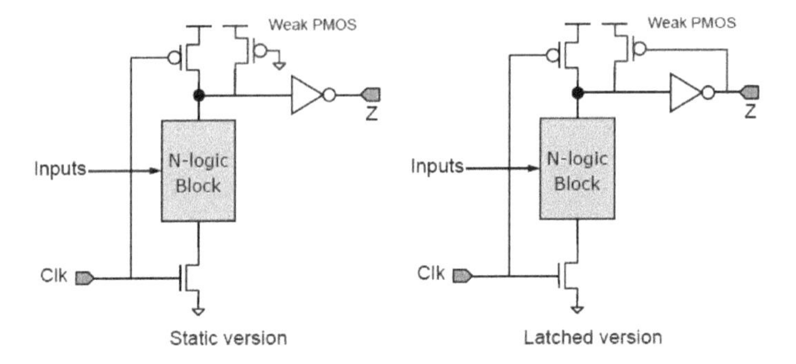

The latched version improves speed and reduces the layout area.

- **Ques: 55. How the limitation of domino logic can be reduced?**

Solution:

The problem of noninversion logic and the static inverter with domino logic can be solved by using alternate logic. That is, the combinational circuits are constructed from alternating NMOS and PMOS blocks. This ensures that the 0 to 1 transition occurs during precharge. It is known as P-E logic.

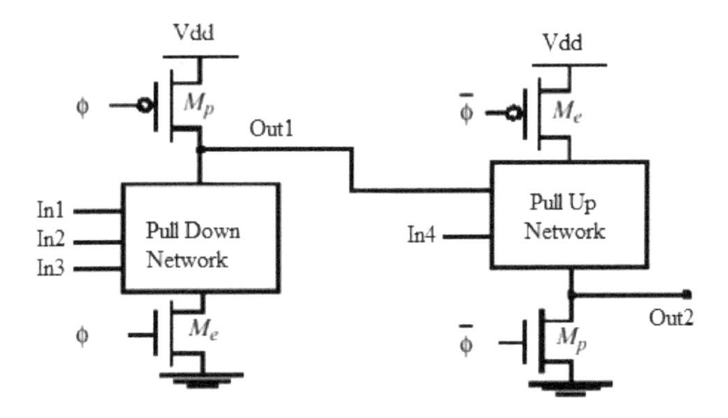

The operation is based on the use of a clock signal that provides two operation phases: the precharge phase and the evaluation phase. The precharge phase is setting the circuit at a predefined initial state while the actual logic response is determined during the evaluation phase.

The precharge output value of φn block equals 1, which is the correct value for the input of a φp block during precharge. All PMOS transistors of the pull-up network (PUN) are turned off; thus, an erroneous discharge at the onset of the evaluation phase is prevented. Similarly, a φn block can follow a

φp block without any problem, as the precharge value of inputs equals 0. To make the evaluation and precharge times of the and φn and φp block coincide, one has to clock the φp block with an invert clock $\varphi\prime$.

The use of the PMOS gates will slow down the logic due to lower mobility of the PMOS. Alternatively, the PMOS transistors have to be larger than the NMOS. The clock also must be routed with its complement. Compared to domino logic, P-E CMOS is more than 20% faster due to the elimination of the static invert and the smaller load capacitance.

- **Ques: 56. What is NORA logic?**

Solution:
NP logic is also known as NORA, i.e., NO RAce.

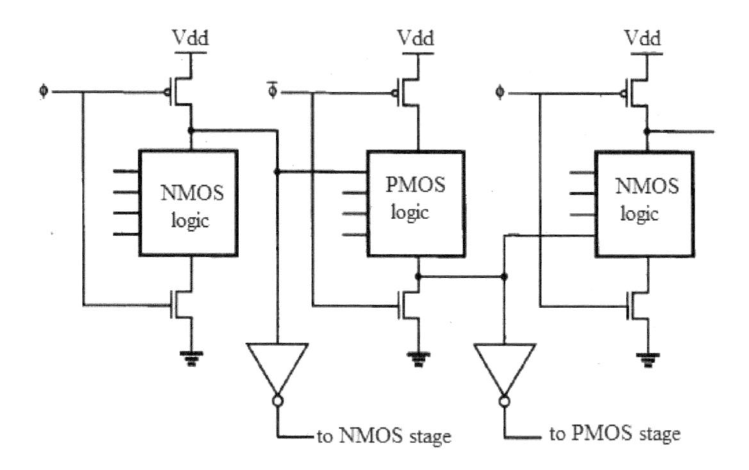

The precharge and evaluate timing of NMOS logic stages is accomplished by the clock signal

Φ, whereas the PMOS logic stages are controlled by the inverted clock signal, ΦT.

When the clock signal is low, the output nodes of NMOS logic blocks are precharged to VDD through the PMOS precharge transistors, whereas the output nodes of PMOS logic blocks are predischarged to 0 V through the NMOS discharge transistors, driven by Φbar.

When the clock signal makes a low-to-high transition (note that the inverted clock signal Φbar makes a high-to-low transition simultaneously), all cascaded NMOS and PMOS logic stages evaluate one after the other, as domino CMOS examined earlier.

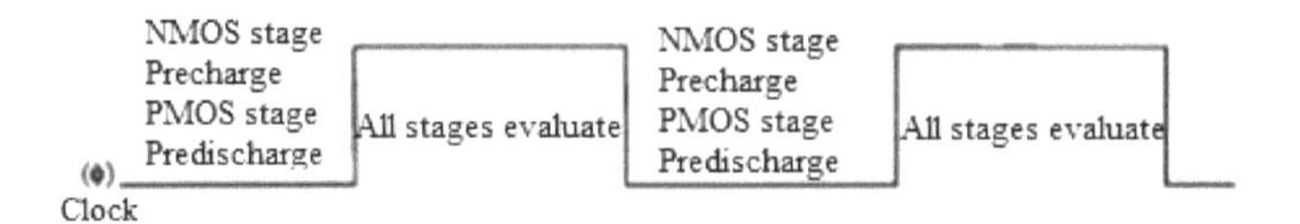

- **Ques: 57. What are zipper CMOS circuits and how are they different from NORA?**

Solution:

The basic circuit architecture of zipper CMOS is essentially identical to NORA CMOS, except the clock signals.

The zipper CMOS clock scheme requires the generation of slightly different clock signals for the precharge (discharge) transistors and the pull-down (pull-up) transistors.

In particular, the clock signals which drive the PMOS precharge and NMOS discharge transistors allow these transistors to remain in weak conduction or near cut-off during the evaluation phase, thus compensating for the charge leakage and charge-sharing problems.

The generalized circuit diagram and the clock signals of the zipper CMOS architecture are shown in the following figure:

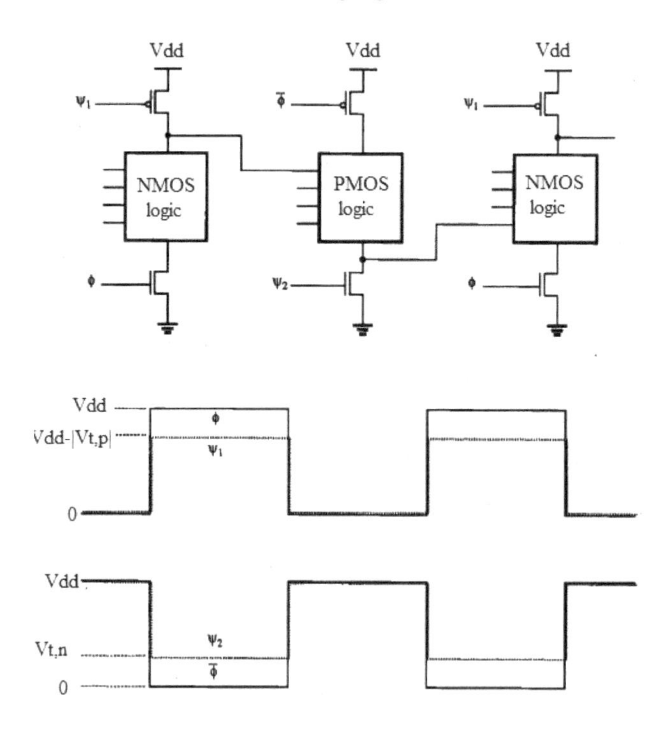

- **Ques: 58. What do you mean by clocked CMOS (C^2MOS)?**

Solution:
Ideally, clocks are nonoverlapping
i.e., $\overline{CLK} \times CLK = 0$
CLK = 1, f is valid
CLK = 0, the output is in a high-impedance state. During this time interval, the output voltage is held on Cout.

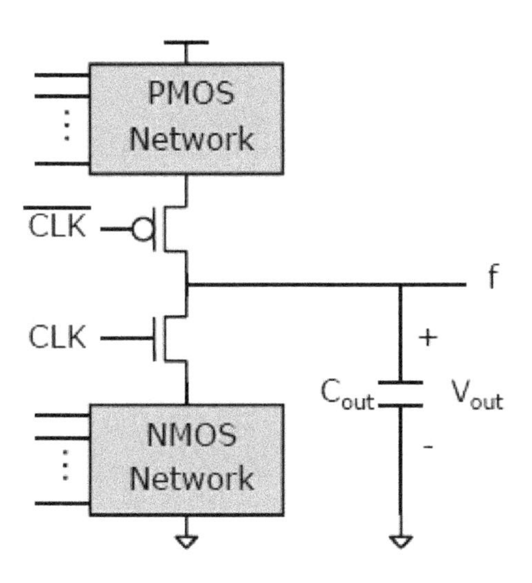

The examples of CCMOS circuits:

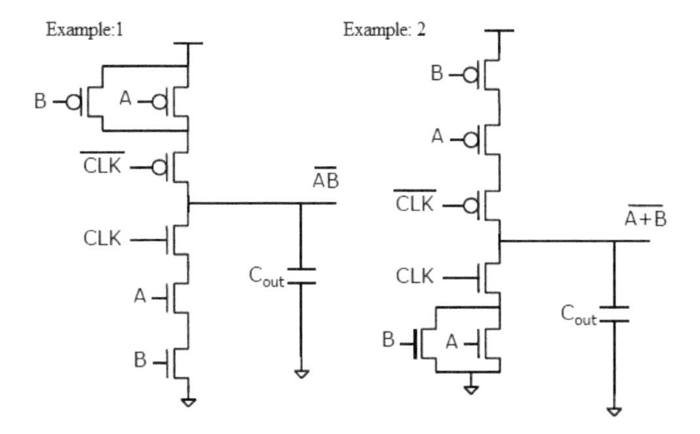

The main drawback is charge sharing, as the output node cannot hold the charge on Vout for a very long duration.
Charge sharing:

$$i_{out} = i_n - i_p = -C_{out}\frac{dV}{dt}$$

$$\Rightarrow dV = -\frac{i_{out}}{C_{out}}dt$$

$$\int_{V_1}^{V(t)} dV = -\int_0^t \frac{I_L}{C_{out}}dt \Rightarrow V(t) = V_1 - \frac{I_L}{C_{out}}t$$

$$V(t_h) = V_1 - \frac{I_L}{C_{out}}t_h = V_X$$

$$t_h = \frac{C_{out}}{I_L}(V_1 - V_X)$$

where it has been assumed that I_{out} is a constant I_L.

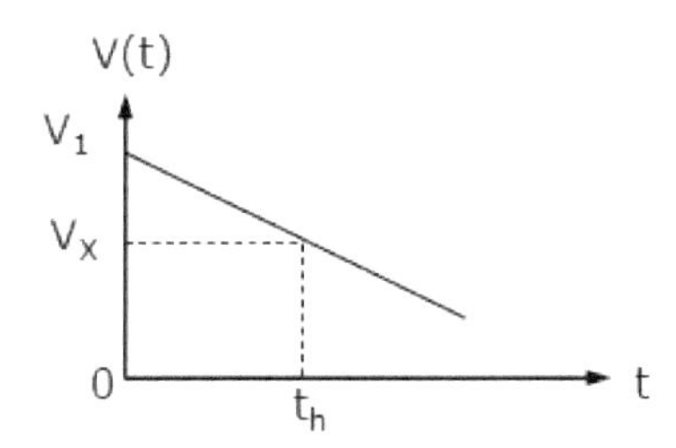

- **Ques: 59. What is adiabatic logic in CMOS?**

Solution:
Adiabatic logic works with the concept of switching activities which reduces the power by giving stored energy back to the supply. Thus, the term adiabatic logic is used in low-power VLSI circuits which implements reversible logic. In this, the main design changes are focused on the power clock which plays a vital role in the principle of operation. Each phase of the power clock gives the user to achieve the two major design rules for the adiabatic circuit design.

The total energy/heat in the system remains constant.
Never turn on a transistor if there is a voltage across it (VDS > 0)
Never turn off a transistor if there is a current through it (IDS $\neq 0$)
Never pass current through a diode

If these conditions about the inputs, in all the four phases of the power clock, the recovery phase will restore the energy to the power clock, resulting in considerable energy saving.

Challenges of energy-recovering circuits:

1. Slowness
2. It requires $\sim$50% off more area than conventional CMOS

- **Ques: 60. Summarize all types of logic used in CMOS.**

Solution:

1. **Domino logic:** It has a minimum area. The power consumption is lower, and the delay is the longest. The DP and AT are average. If the design goal is minimum area and number of the transistors and the speed can be sacrificed a bit, then domino logic is the best structure for Ripple Carry Adder.
2. **P-E logic:** It has a smaller area. The power consumption is lower, and the delay is shorter.
   It has the lower DP and AT for Ripple Carry Adder. If the logic has no inherent race problem, it will be the better choice for Ripple Carry Adder.
3. **P-E (race-free) logic:** To avoid the race condition of P-E Logic, the P-E (race-free) logic is introduced. It has a small area. The area is larger than P-E logic. The power consumption is average. The delay is shorter. It has a lower DP and AT for Ripple Carry Adder. For synthesis, it is the better choice for Ripple Carry Adder.
4. **NORA logic:** Power consumption is the highest. The area is small. The delay is longer. The DP is highest and AT is higher. The performance of this dynamic logic is so average.
5. **2-Phase logic**
   (1) P-type 2-Phase logic: The area is small. The delay is shortest in the Ripple Carry structure.
   The power consumption is lower than other logic structures. DP is lower and AT is high. If the design goal is high speed, this logic structure for designing Ripple Carry Adder is a good choice.

(2) N-type 2-Phase logic: The area is small. The delay is average. The power consumption is the lowest. DP and AT are the lowest than other logic structures of Ripple Carry Adders. If the design goal is DP and AT, this structure is the best choice.

(3) Domino 2-Phase logic: The area is large. The delay is short. The power consumption is lower. DP is lower and AT is the same as P-E logic. The performance of this logic is good except for having a large area.

6. **MODL logic**: The adder that has been designed by this logic is Carry-Look-ahead Adder. So, compared with Ripple Carry Adder, Carry-Look-ahead Adder has better performance of delays, DP and AT. Especially, the delay and AT are 60% and 80% less then Ripple Carry Adder.
7. **Cascode logic**: It has lower power consumption and the largest area. The delay is shorter. DP is less than other Ripple Carry Adders except for n-type 2-Phase logic. But AT is the highest. So, if the design goal is high speed and low power with little attention to the area then Cascode is a better choice.

- **Ques: 61. In MOSFET, the gate-body connection exhibits many properties of a capacitor. Justify it.**

Solution:
The gate terminal of a MOSFET is typically considered its input. Since the gate is separated from the semiconductor body by an insulating oxide layer, the gate-body connection exhibits many properties of a capacitor.

1. One property is that no current flows into the gate terminal on average.
2. Another capacitive property is that the gate and body terminals can support an electric field between them. This electric field can attract mobile charges toward either the gate or body terminals and repel them from the other. By changing the charge distribution between gate and body, the MOSFET's drain-source (or source-drain) connection can be opened or connected. In this way, NMOS and PMOS transistors operate as a switch.

- **Ques: 62. What is PTL (pass transistor logic)?**

Solution:
Pass-transistor logic (PTL), also known as transmission-gate logic, is based on the use of MOSFETs as switches rather than as inverters.

The pass transistor logic is required to reduce the transistors for implementing logic by using the primary inputs to drive gate terminals, source, and drain terminals. In complementary CMOS logic, primary inputs are allowed to drive only gate terminals.

The figure that follows shows the implementation of AND function using the only NMOS pass transistors.

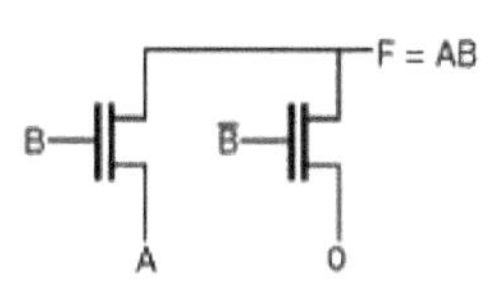

A	B	F
0	0	0
0	1	0
1	0	0
1	1	1

In this gate, if the B input is high the left NMOS is turned ON and copies the input A to the output F. When B is low the right NMOS pass transistor is turned ON and passes a “0” to the output F.

The major advantages of pass transistor logic are as follows:

1. Fewer transistors are required to implement a given function. To illustrate this, consider the implementation of AND gate using complementary CMOS logic. If we compare this with the same AND gate implementation using pass transistor logic the number of transistors required is four including the two-transistor required to invert the input B.
2. Lower capacitance because of the reduced number of transistors. As discussed NMOS devices are effective in passing strong “0” but it is poor at pulling a node to VDD. Hence when the pass transistor pulls a node to high logic the output only changes up to VDD–Vth. This is the major disadvantage of pass transistors.

- **Ques: 63. What do you understand by latch-up?**

Solution:
Latch-up is a functional chip failure associated with excessive current going through the chip, caused by weak circuit design. In some cases, latch-up can be a temporary condition that can be resolved by a power cycle, but unfortunately, it can also cause fatal chip failure.

CMOS Latch-Up

The following diagram is a drawing of a typical CMOS circuit. Notice the two transistors, NPN and PNP and their connection to VDD and GND supply rails. The two transistors are protected by resistors but if examined more closely, there's an SCR device that could be triggered.

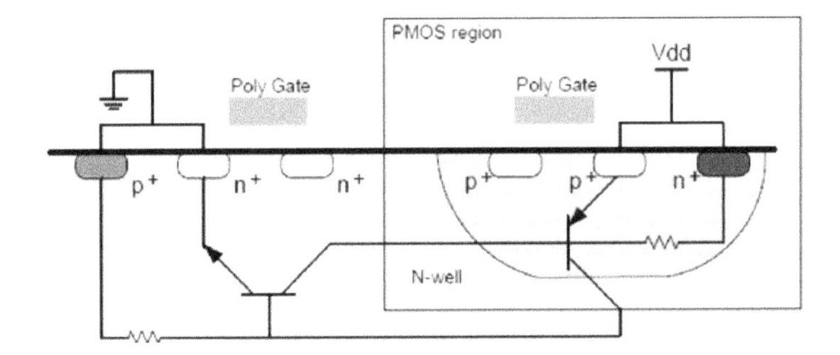

In latch-up conduction, the current flows from VDD to GND directly via the two transistors, causing the dangerous condition of a short circuit. The resistors are bypassed and thus excessive current flows from VDD to ground.

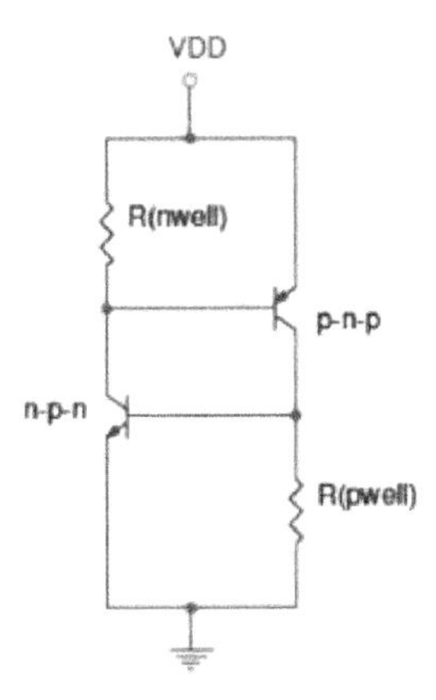

So, latch-up is a state where a semiconductor undergoes a high-current state (or low impedance path) as a result of the interaction of PNP and NPN bipolar transistors.

In CMOS, these transistors act as parasitic elements. When an interaction occurs between PNP and NPN transistors, regenerative feedback between the two transistors leads to electrical instability.

When a low impedance path is created between supply and ground, the short circuit between power and ground rails in ICs leads to high current and damage to IC.

This problem is considered because it initiates the thermal runaway and thereafter leading to the destruction of the semiconductor. To avoid the latch-up problem NMOS and PMOS devices are surrounded by an oxide layer. This provides a high resistance and limits the large current flow.

- **Ques: 64. What is the significance of ESD? Explain ESD using models.**

Solution:
Electrostatic discharge (ESD) is a charge balancing process between two objects at a different potential. The phenomenon of ESD can often be observed in our daily lives. For example, static electricity can be generated due to the friction between different materials, and the accumulated electrostatic charge can spontaneously be transferred to the object at lower potential; either through direct contact or through an induced electric field, just as when you reach a doorknob on one dry day. ESD events usually give a mild shock to human beings. ESD events often involve high voltage (˜several kV) and high current stress (1–10 A) on small electric devices. Even though ESD events are of very short duration (0.2–200 ns), the massive current/voltage pulses can effect fatal damage to ICs. There are several examples of catastrophic ESD failures in modern ICs, such as junction breakdown, molten metal/via effects, and the gate oxide damage. Besides these noticeable failures, minor levels of ESD shock can also generate latent problems within the devices, producing the so-called "walking wounded" devices.

One of the most frequently observed ESD events is the transfer of electrostatic charge from a charged human body to an ESD sensitive device due to improper handling. The model developed to represent this event is the Human Body Model (HBM), which is the most classical and common industrial test method. In the HBM, it is assumed that a certain amount of electrostatic charge initially is stored on the body and the charge is transferred to an object through a finger when the physical contact between the charged human body and the object is made.

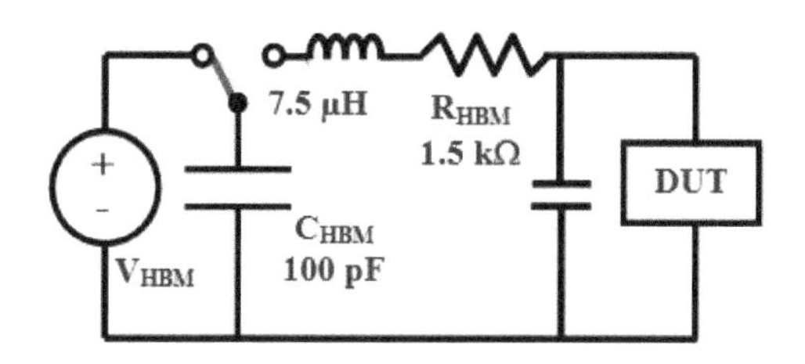

Human Body Model

It is a simplified equivalent circuit of HBM ESD conditions. It consists of a charging capacitor, and contact resistance between the charge source and the Device Under Test (DUT). In the HBM standard, the circuit component used to simulate the charged human body is a 100 pF capacitor and the resistance of the discharging path is 1500 Ω; it electrically looks like a current source if the DUT provides a current path of low resistivity. LHBM (~0.75 μH) is the effective inductance of the discharge path in a real tester. HBM has the longest pulse among the three primary ESD models. The rise time of the HBM pulse is approximately 5–10 ns, and the decay time is ~150 ns[6].

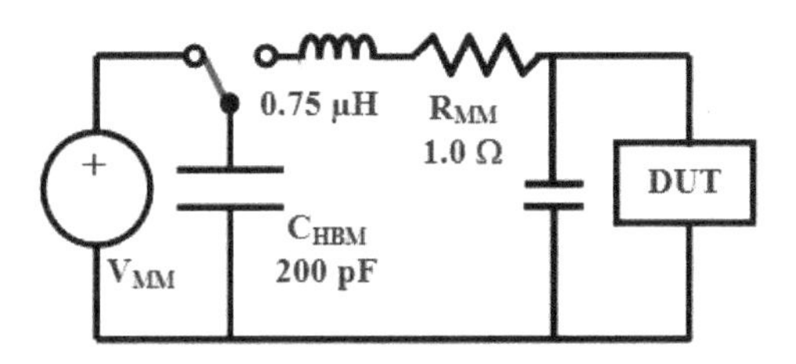

Machine Model

- **Ques: 65. Explain the ESD protection circuit.**

Solution:

ESD-protection circuits were first integrated into CMOS devices. The thin and, therefore, very vulnerable gate oxide of the MOS transistor makes protection against destruction as a result of ESDs essential. The protective precaution that was taken initially, and which is still the best method, is the integration of clamping diodes, which limit the dangerous voltages and conduct excess currents into regions of the circuit that are safe. The safe regions consist primarily of supply-voltage connections.

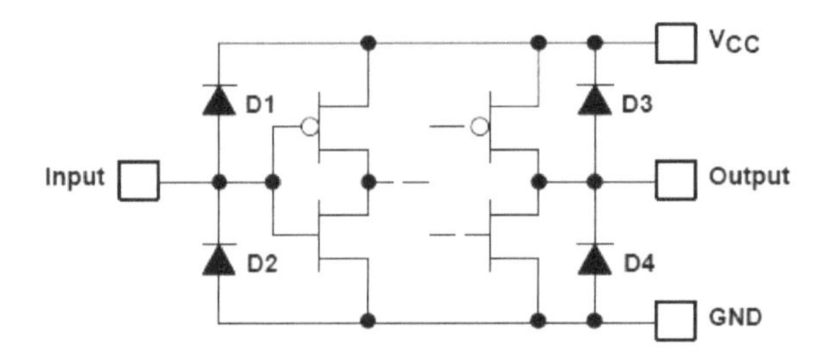

In the simplest case, the protection circuits consist of diodes that are oriented to be blocking in normal operation and are situated between the connection to the component to be protected and the supply voltage lines.

- **Ques: 66. Explain the CMOS layout of the tristate buffer.**

Solution:
The symbol of the tristate buffer is:

in — en — out

CMOS layout of the tristate buffer:

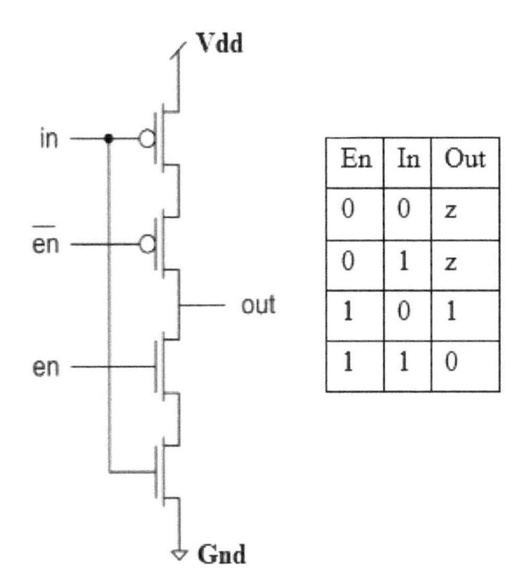

En	In	Out
0	0	z
0	1	z
1	0	1
1	1	0

When En = 0,
The PMOS part of the circuit, as well as NMOS part of the circuit, doesn't conduct, therefore no one drives the output, resulting in a high impedance circuit.
When En = 1; In = 0,
The PMOS part of the circuit conducts resulting in "1" at the output.
When En = 1; In = 1,
The NMOS part of the circuit conducts resulting in "0" at the output.

- **Ques: 67. Explain the transmission gate.**

Solution:
A transmission gate, or analog switch, is defined as an electronic element that will selectively block or pass a signal level from the input to the output. This switch is comprised of a PMOS transistor and an NMOS transistor. The control gates are biased in a complementary manner so that both transistors are either on or off.

When the voltage on node X is a "1," the complementary "0" is applied to node active-low X, allowing both transistors to conduct and pass the signal at A to B. Similarly, when the voltage on node active-low X is a "0," the complementary "1" is applied to node A, turning both transistors off and forcing a high-impedance condition on both A and B nodes.

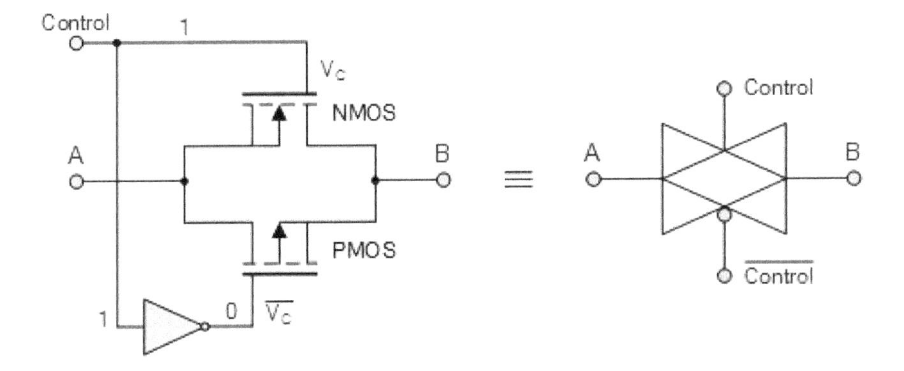

The labels A and B can be reversed.

- **Ques: 68. Explain 2:1 multiplexer using transmission gate.**

Solution:
2:1 MUX using transmission gate:

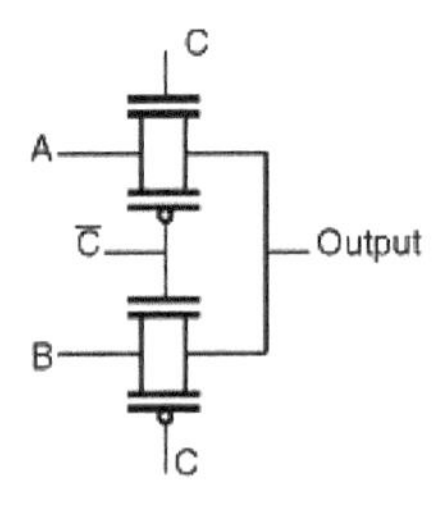

When the control signal C is high then the upper transmission gate is ON and it passes A through it so that output = A.

When the control signal C is low then the upper transmission gate turns OFF and it will not allow A to pass through it, at the same time the lower transmission gate is "ON" and it allows B to pass through it so the output = B.

- **Ques: 69. What are the benefits of CMOS logic?**

Solution:

- ☐ Extremely large fan-out (>50)
- ☐ Lowest power dissipation
- ☐ Very high noise immunity and noise margin
- ☐ Lower propagation delay than NMOS
- ☐ Higher speed than NMOS (about 4 GHz)
- ☐ Large logic swing (i.e., equal to Vdd)

- **Ques: 70. Enlist any three disadvantages of CMOS logic.**

Solution:

- ☐ It is costly
- ☐ Packing density is less
- ☐ Static charges

- **Ques: 71. Why threshold is negative for PMOS?**

Solution:
To create a p-type channel in PMOS, it is required to attract holes toward the gate. Therefore, Vgs is negative, as holes are attracted to a negative voltage.

- **Ques: 72. How CMOS is different than TTL?**

Solution:
The CMOS is different than TTL (transistor–transistor logic) in the following ways:

1. The CMOS uses field-effect transistors, whereas the TTL utilizes the BJT.
2. The CMOS draws lesser power than the TTL. A single gate in the CMOS chip can consume around 10 nW while an equivalent gate on the TTL chip can consume around 10 mW of power.
3. The TTL needs extra components such as resistors, whereas the CMOS consists of only the FETs.
4. The CMOS are more susceptible to electrostatic discharge.

5. The basic gates used in standard TTL are NAND gates while NAND-NOR gates are used in the CMOS circuits.
6. Fan-out for the TTL is 10 while it is 50 for the CMOS.
7. The propagation delay for the TTL is 10 ns and for the CMOS it is 70 ns.
8. Fan-in for the TTL is ∼12–14 and it is >10 for the CMOS.
9. For the TTL, the noise margin is 0.5 V while for the CMOS it is 1.5 V.
10. Noise immunity of the CMOS is a lot better than the TTL circuits.
11. The CMOS circuits are simpler to construct and have a higher packing density than the TTL logic family.

References

1. Uyemura, J. P. (2012). *Circuit Design for CMOS VLSI.* Springer Science & Business Media.
2. Weste, N. H. and Harris, D. (2015). *CMOS VLSI Design: Circuits and Systems Perspective*. Pearson Education India.
3. https://www.elprocus.com/cmos-working-principle-and-applications/
4. Pucknell, D. A. and Eshraghian, K. (1994). *Basic VLSI Design*. Prentice-Hall of India.
5. Geiger, R. L., Allen, P. E., and Strader, N. R. (1990). VLSI design techniques for analog and digital circuits.

4

Verilog Hardware Descriptive Language

Hardware description language (HDL) is a specialized computer language used to program electronic and digital logic circuits. The structure, operation, and design of the circuits are programmable using HDL. Further, HDL includes a textual description consisting of operators, expressions, statements, inputs, and outputs.

Instead of generating a computer-executable file, the HDL compilers provide a gate map. The gate map obtained is then downloaded to the programming device to check the operations of the desired circuit. The language helps to describe any digital circuit in the form of structural, behavioral, and gate-level and it is found to be an excellent programming language for field programmable gate arrays (FPGAs) and complex programmable logic devices (CPLDs).

4.1 Commonly Used HDL

The HDLs will allow fast design and better verification. The three commonly used HDLs are the following:

- Verilog HDL
- VHDL (very high-speed integrated circuit hardware description language).
- SystemC

In most of the industries, Verilog and VHDL are common. Verilog, one of the main HDL standardized as IEEE 1364 is used for designing all types of circuits. It consists of modules and the language allows behavioral, dataflow, and structural description. VHDL is standardized by IEEE1164. The design is composed of entities consisting of multiple architectures. SystemC is a language that consists of a set of C++ classes and macros. It allows electronic system level and transaction modeling.

4.2 Need For HDL

Moore's law in the year 1970 has brought a drastic change in the field of integrated circuit (IC) technology. This change has made developers bring out complex digital and electronic circuits. But the problem was the absence of a better programming language allowing hardware and software codesign. Complex digital circuit designs require more time for development, synthesis, simulation, and debugging. The arrival of HDLs has helped to solve this problem by allowing each module to be worked by a separate team.

All the goals such as power, throughput, latency (delay), test coverage, functionality, and area consumption required for a design can be known by using HDL. As a result, the designer can make the necessary engineering tradeoffs and can develop the design in a better and efficient way. Simple syntax, expressions, statements, concurrent, and sequential programming are also necessary while describing the electronic circuits. All these features can be obtained by using an HDL. Now while comparing HDL and C languages, the major difference is that HDL provides the timing information of a design.

4.3 HDL Simulation and Debugging

HDL simulation, which involves the need for a test bench. A test bench is an environment used to check the correctness of the design. The test bench involves the inputs, outputs, and the procedure to be done. The HDL simulator will execute the test bench; hence, the designer can verify the function of the circuit on the waveform analyzer provided by the HDL simulator. Many simulators are there for HDLs.

Debugging can be done before and after simulation. If errors are present, the HDL simulators usually give an error report and thereby corrections can be done. Also, by checking the simulation result the errors can be found and rectified. ModelSim, Xilinx ISim, and Aldec Active-HDL are some examples of HDL simulators.

4.4 Different Types of HDL

Different HDLs are available for describing analog circuits, digital circuits, and PCBs.

1. HDLs for digital circuit design
   Other than Verilog, VHDL, and SystemC many HDLs are available for digital circuits.

- Advanced Boolean Expression Language (ABEL) is better for programming PLDs (programmable logic devices). It includes sequential and concurrent logic formats, truth table formats, and describes test vectors also.
- Altera hardware description language (AHDL), a language from Altera is used for CPLDs and FPGAs. All language constructs can be used in AHDL providing more control and support.
- Bluespec, a high-level functional programming HDL language is developed to handle chip design and automation systems. Bluespec System Verilog (BSV) uses a syntax similar to Verilog HDL.
- C-to-Verilog is generally a converter that helps to convert C to Verilog language. Constructing hardware in a Scala Embedded Language (Chisel), MyHDL, and HHDL is HDLs used to support advanced hardware design.
- Compiler for Universal Programming Language (CUPL) is generally used for logic device programming.
- Handel-C is used for programming FPGAs. Hardware Join Java (HJJ) helps in reconfigurable computing applications.
- Impulse C, a subset of C supports parallel programming.
- Just-another hardware description language (JHDL) uses an object-oriented programming approach. LavaHDL mainly helps in specifying the layout of circuits.
- Lola HDL is used for synchronous digital circuits.
- M is another HDL from Mentor Graphics.
- PALASM is used as an HDL for programmable array logic (PAL) devices.
- Finally, System Verilog is an extension of Verilog.

2. HDLs for analog circuit design:
   The HDLs used for analog circuits include the following:
   - Analog HDL(AHDL)
   - Spectre high-level description language (SpectreHDL)
   - Verilog for analog and mixed signal (Verilog-AMS)
   - VHDL with analog and mixed-signal extension (VHDL-AMS)
   - HDL-A is a proprietary HDL for mixed and analog circuits.
   - AHDL is most commonly used as an HDL language for analog circuits.
   - SpectreHDL is a high-level description language that uses functional description text files to model the behavior of the systems.

- Verilog-AMS is an industry standard modeling language that contains continuous, an event-driven simulator for analog, digital, and analog/digital circuits.
- VHDL-AMS is good for verifying complex analog, mixed-signal, and radio frequency (RF) circuits.

3. HDL for printed circuit design
   PHDL (printed circuit board HDL) is generally used for modeling text-based schematics for PCBs. It allows for generating massive buses and reuse device definitions easily.

Benefits of HDL

The major benefit of the language is a fast design and better verification. The top-down and hierarchical design methods allow the design time; design cost; and design errors to be reduced. Another major advantage is related to complex designs, which can be managed and verified easily. HDL provides the timing information and allows the design to be described at the gate level and register transfer level. The reusability of resources is one of the other advantages.

Technical Questions with Solutions

- **Ques: 1. What do you mean by HDL?**

Solution:
The computer language which is used to program electronic and digital logic circuits is known as hardware description language (HDL). The HDL is used to program structure, operation, and design of circuits. It represents the behavior of digital circuits.

HDL includes a textual description consisting of operators, expressions, statements, inputs, and outputs. Instead of generating a computer-executable file, the HDL compilers provide a gate map. The gate map obtained is then downloaded to the programming device to check the operations of the desired circuit.

The language helps to describe any digital circuit in the form of structural, behavioral, and gate-level and it is found to be an excellent programming language for FPGAs and CPLDs.

- **Ques: 2. Name any six types of HDL languages.**

Solution:

1. VHDL (VHSIC HDL)
2. Verilog
3. System C
4. System Verilog
5. Advanced Boolean Expression Language (ABEL)
6. Altera Hardware Description Language (AHDL)

- **Ques: 3. What is the need of HDL?**

Solution:
Moore's law in the year 1970 has brought a drastic change in the field of integrated circuit (IC) technology. This change has made the developers bring out complex digital and electronic circuits.

1. Decades back, designers designed their system with the help of truth tables, Boolean mathematics, K-maps, and other expression-solving techniques. But chip density and complexity were increasing continuously and then thousands of gates in a single chip were common and designing that system was not an easy task for engineers. So, HDL emerged for designing and verification of circuits.

2. Earlier, the designer and verification engineers worked on the printed circuit boards or bread boards for test and verify their designs, but that was not the practical approach to verify designs because that takes too much time and also the verification engineers were not sure about the verification. Each gate must be checked before using logical verification which was not practically possible. Testing of large circuits was not possible because the large circuits contain lots of gates and the response of every gate could not be checked. So, HDL is the best option for logical verification.

3. Earlier, the designing of circuits requires more manpower as well as time. After using of HDL, the designing time is less because HDL is ready to work on sequential execution of statements and the designers only have to write their requirements in the term of language and then synthesis tools convert that high-level language into structural or RTL level or gate-level design.

4. Hence, using HDL's circuit designing is very easy through computer and then we can also develop these codes on hardware using PLDs. and it saves time and money.

- **Ques: 4. Which IEEE standard is used in VHDL and Verilog for designing all types of circuits?**

Solution:
Verilog, one of the main HDL standardized as IEEE 1364 is used for designing all types of circuits. It consists of modules and the language allows behavioral, dataflow, and structural description.

VHDL (very high speed integrated circuit hardware description language) is standardized by IEEE1164. The design is composed of entities consisting of multiple architectures.

- **Ques: 5. Which HDLs are used to design digital circuits?**

Solution:
Other than Verilog, VHDL and SystemC many HDLs are available for digital circuits.

1. ABEL is better for programming PLDs (programmable logic devices). It includes sequential and concurrent logic formats, truth table formats, and describes test vectors also.
2. AHDL, a language from Altera is used for CPLDs and FPGAs. All language constructs can be used in AHDL providing more control and support.
3. Bluespec, a high-level functional programming HDL language is developed to handle chip design and automation systems. Bluespec System Verilog (BSV) uses a syntax similar to Verilog HDL.
4. C-to-Verilog is generally a converter that helps to convert C to Verilog language. Constructing hardware in a Scala Embedded Language (Chisel), MyHDL, and HHDL are HDLs used to support advanced hardware design.
5. Confluence and CoWareC are the two languages that were discontinued after the arrival of the System C language.
6. Compiler for Universal Programming Language (CUPL) is generally used for logic device programming.
7. Handel-C is used for programming FPGAs.
8. Hardware Join Java (HJJ) helps in reconfigurable computing applications.
9. Impulse C, a subset of C supports parallel programming.
10. Just-Another Hardware Description Language (JHDL) uses an object-oriented programming approach.
11. Lava HDL helps in specifying the layout of circuits mainly.
12. Lola HDL is used for synchronous digital circuits.
13. M is another HDL from Mentor Graphics.

14. PALASM is used as an HDL for Programmable Array Logic (PAL) devices.
15. Finally, System Verilog is an extension of Verilog.

- **Ques: 6. Which HDLs are used to design analog circuits?**

Solution:
The HDLs used for analog circuits include:

1. Analog Hardware Description Language (AHDL)
   AHDL is most commonly used as an HDL language for analog circuits.
2. Spectre High-Level Description Language (SpectreHDL)
   SpectreHDL is a high-level description language that uses functional description text files to model the behavior of the systems.
3. Verilog for analog and mixed signal (Verilog-AMS)
   Verilog-AMS is an industry-standard modeling language that contains continuous, an event-driven simulator for analog, digital, and analog/digital circuits.
4. VHDL with analog and mixed-signal extension (VHDL-AMS)
   VHDL-AMS is good for verifying complex analog, mixed-signal, and RF circuits.
5. HDL-A
   HDL-A is a proprietary HDL for mixed and analog circuits.

- **Ques: 7. Which HDLs are used to design PCBs?**

Solution:
PHDL (printed circuit board HDL) is generally used for modeling text-based schematics for PCBs. It allows for generating massive buses and reuse device definitions easily.

- **Ques: 8. What are the benefits of HDL?**

Solution:

1. Fast design and better verification.
2. The top-down design and hierarchical design method allow the design time; design cost; and design errors to be reduced.
3. A complex design can be managed and verified easily.
4. HDL provides the timing information and allows the design to be described at the gate level and register transfer level.
5. The reusability of resources is one of the other advantages.

- **Ques: 9. What are the different steps in the VLSI IC design flow?**

Solution:
Logic Synthesis

- RTL conversion into a netlist
- Design partitioning into physical blocks
- Timing margin and timing constraints
- RTL and gate-level netlist verification
- Static timing analysis

Floor planning

- Hierarchical VLSI blocks placement
- Power and clock planning

Synthesis

- Timing constraints and optimization
- Static timing analysis
- Update placement
- Update power and clock planning

Placement and Routing

- Complete placement and routing of blocks

VLSI Level Layout

- VLSI integration of all blocks
- Place and route
- GDSII creation (used by the foundry to fabricate the silicon)

- **Ques: 10. What do you mean by RTL?**

Solution:
RTL is used in the logic design phase of the IC design cycle. RTL design or register transfer level design is a method in which we can transfer data from one register to another.

An RTL description is usually converted to a gate-level description of the circuit by a logic synthesis tool. The synthesis results are then used by placement and routing tools to create a physical layout.

RTL is based on synchronous logic and contains three constraints, registers which hold state information, combinatorial logic which defines the next state inputs and clocks that control when the state changes.

- **Ques: 11. What do you mean by netlist?**

Solution:
The netlist is a textual description of all the nets present in your design. It can be the connectivity of different components like gates (generally in gate-level netlist as all the components are gates), resistors, capacitors, inductors, or transistors. After drawing a circuit, the program can generate a “netlist” which is a list of all components used in the circuit.

Gate-level netlist means design synthesized into gates and flip flops. Gate-level netlist is nothing but interconnections of logic blocks and logic cells.

RTL netlist means RTL code which is not synthesized

Transistor netlist means to design in terms of transistors.

- **Ques: 12. What is the difference between RTL and netlist?**

Solution:

RTL:

RTL simply means register transfer logic. As the expansion says it means data is transferred between registers/flops. For example, if in a design we want to communicate between two blocks from B1 to B2. Here we transfer data from B1 to a flipflop and then to B2. This way of interfacing is referred to as RTL. The functionality of a device is written in a language like Verilog, VHDL. It is called RTL if it can be synthesized, that is, it can be converted to the gate-level description.

RTL uses a combination of behavioral and dataflow constructs and is acceptable to logic synthesis tools.

Netlist

The netlist is achieved after synthesizing RTL. This is a gate-level description of the device. The high-level statements in RTL will be converted to the gate level. For timing, the simulation netlist is used with SDF (contains delay information). Netlist refers to the actual implementation of a particular logic or design and its interconnections.

- **Ques: 13. How many digital design methodologies are used in Verilog HDL?**

Solution:
There are two basic types of digital design methodologies:

1. Top-down design methodology
2. Bottom-up design methodology

In a top-down design methodology, we define the top-level block and identify the sub-blocks necessary to build the top-level block. The sub-blocks are further subdivided into leaf cells, which are the cells that cannot further be divided.

In a bottom-up design methodology, identify the leaf cells and create the bigger cells, which are further used for higher-level blocks until we build the top-level block in the design.

- **Ques: 14. What do you understand by "module" in Verilog HDL?**

Solution:
A module is the basic building block in Verilog. A module can be an element or a collection of lower-level design blocks. A module is the principal design entity in Verilog.

The first line of a module declaration specifies the name and port list (arguments). The next few lines specify the i/o type (input, output or inout) and width of each port.

Syntax

```
module module_name (port_list);
  input [msb:lsb] input_port_list;
  output [msb:lsb] output_port_list;
  inout [msb:lsb] inout_port_list;
     ... statements ...
end module
```

Typically, the inputs are wire since their data is latched outside the module. Outputs are type reg if their signals were stored inside an always or initial block.

One module definition **cannot** contain another module definition within the module and endmodule statements.

- **Ques: 15. What does the level of abstraction mean?**

Solution:
The level of abstraction is the amount of complexity by which a system is viewed or programmed. The higher the level, the less the detail. The lower the level, the more the detail. The highest level of abstraction is the entire system. The next level would be a handful of components, and so on, while the lowest level could be millions of components.

- **Ques: 16. In Verilog HDL, how many "levels of abstractions" are there?**

Solution:

1. The behavioral or algorithmic level

This is the highest level of abstraction provided by Verilog HDL. A module can be implemented in terms of the desired design algorithm without concern for the hardware implementation details. Designing at this level is very similar to C programming.

2. Dataflow level

At this level, the module is designed by specifying the data flow. The designer is aware of how data flows between hardware registers and how the data is processed in the design.

3. Gate level

The module is implemented in terms of logic gates and interconnections between these gates. Design at this level is similar to describing design in terms of a gate-level logic diagram.

4. Switch level

This is the lowest level of abstraction provided by Verilog. A module can be implemented in terms of switches, storage nodes, and the interconnections between them. Design at this level requires knowledge of switch-level implementation details.

- **Ques: 17. In Verilog, name the highest and lowest level of abstraction.**

Solution:

The behavioral or algorithmic level is the highest level of abstraction provided by Verilog HDL.

Switch level is the lowest level of abstraction provided by Verilog HDL.

The higher the level of abstraction, the more flexible and technology-independent the design. As one goes lower toward switch-level design, the design becomes technology-dependent and inflexible.

- **Ques: 18. What is the module instantiation?**

Solution:

The design hierarchy by instantiating modules in other modules. A module can be instanced when that module is used in another, higher-level module.

```
module dff (clk, d, q);
input clk, d;
output q;
reg q;
always @(posedge clk) q = d;
endmodule
```

```
module top;
reg data, clock;
wire q_out, net_1;
  dff inst_1 (.d(data), .q(net_1), .clk(clock));
  dff inst_2 (.clk(clock), .d(net_1), .q(q_out));
endmodule
```

The process of creating objects from a module template is called instantiation, and the objects are called instances.

- **Ques: 19. What is "port" in Verilog HDL?**

Solution:
Ports allow communication between a module and its environment. All but the top-level modules in a hierarchy have ports. Ports can be associated by order or by name.

Example

```
input          clk; // clock input
input [15:0]   data_in; // 16-bit data input bus
output [7:0]   count; // 8-bit counter output
inout          data_bi; // Bi-Directional data bus
```

- **Ques: 20. What are the port connection rules?**

Solution:

- Inputs: internally must always be of type net, externally the inputs can be connected to a variable of type reg or net.
- Outputs: internally can be of type net or reg, externally the outputs must be connected to a variable of type net.
- Inouts: internally or externally must always be type net, can only be connected to a variable net type.
- Width matching: It is legal to connect internal and external ports of different sizes. But beware, synthesis tools could report problems.
- Unconnected ports: unconnected ports are allowed by using a ",".
- The net data types are used to connect structure.
- A net data type is required if a signal can be driven by a structural connection.

- **Ques: 21. What do you understand by "stimulus"?**

Solution:
Once a design block is completed, it must be tested. The functionality of the design block can be tested by applying stimulus and checking results. We

call such a block the stimulus block. It is a good practice to keep the stimulus and design blocks separate. The stimulus block can be written in Verilog. A separate language is not required to describe stimulus. The stimulus block is also commonly called a test bench. Different test benches can be used to thoroughly test the design block.

- **Ques: 22. What do you mean by lexical convention in Verilog?**

Solution:
The lexical conventions provide the necessary framework for Verilog HDL. Verilog source text files consist of the following lexical tokens:

- Whitespace
- Comments
- Operators
- Number specification
- Strings
- Identifiers and keywords
- Escaped identifiers

- **Ques: 23. How numbers are stored in Verilog?**

Solution:

Sized numbers

The sized numbers are represented as: <size> '<base format> <number>
<size> is written only in decimal and specifies the number of bits in the number. Legal base formats are decimal ('d or 'D), hexadecimal ('h or 'H), binary ('b or 'B) and octal ('o or 'O).
e.g. 3'b111 // This is a 3-bit binary number

Unsized numbers

The unsized numbers that are specified without a <base format> specification are decimal numbers by default. Numbers that are written without a <size> specification have a default number of bits that is a simulator- and machine-specific.

e.g. 'hd2 // This is a 32-bit hexadecimal number

X or Z numbers

An unknown value is denoted by an x. A high impedance value is denoted by z.

e.g., 12’h28x // This is a 12-bit hex number; 4 least significant bits unknown

Negative numbers

Negative numbers can be specified by putting a minus sign before the size for a constant number. Size constants are always positive. It is **illegal** to have a minus sign between <base format> and <number>. An optional signed specifier can be added for signed arithmetic.

e.g. 4’d-2 // Illegal specification

- **Ques: 24. What do you mean by “strings” in Verilog?**

Solution:
A string is a sequence of characters that are enclosed by double quotes. The restriction on a string is that it must be contained on a single line, that is, without a carriage return. It cannot be on multiple lines. Strings are treated as a sequence of one-byte ASCII values.

"Hello Verilog World" // is a string

- **Ques: 25. What are “keywords and identifiers” in Verilog?**

Solution:
Keywords are special identifiers reserved to define the language constructs. Keywords are in lowercase. These keywords have special meaning in Verilog.

Identifiers are names given to objects so that they can be referenced in the design. Identifiers are made up of alphanumeric characters, the underscore (_), or the dollar sign ($). Identifiers are case sensitive. Identifiers start with an alphabetic character or an underscore. They cannot start with a digit or a $ sign.

e.g., reg value; // reg is a keyword; value is an identifier

- **Ques: 26. what is “escaped identifiers” in Verilog?**

Solution:
Escaped identifiers begin with the backslash (\) character and end with whitespace (space, tab, or newline). All characters between backslash and whitespace are processed literally. Any printable ASCII character can be included in escaped identifiers. Neither the backslash nor the terminating whitespace is considered to be a part of the identifier.

e.g., \x+y-z

- **Ques: 27. What are the various data values in Verilog?**

Solution:
There are four types of data values:

1: True condition
0: False condition
X: Unknown number
Z: High impedance

- **Ques: 28. What is the importance of "net" in Verilog?**

Solution:
Nets represent connections between hardware elements. Nets are declared primarily with the keyword wire. Nets are one-bit values by default unless they are declared explicitly as vectors. The terms wire and net are often used interchangeably. The default value of a net is z (except the trireg net, which defaults to x). Nets get the output value of their drivers. If a net has no driver, it gets the value z.

- **Ques: 29. Write a Verilog code to swap the contents of two registers with and without a temporary register?**

Solution:
With temp reg;

```
always @ (posedge clock)
begin
temp=y;
y=x;
x=temp;
end
```

Without temp reg;

```
always @ (posedge clock)
begin
x <= y;
y <= x;
end
```

- **Ques: 30. What is the difference between blocking and nonblocking?**

Solution:
The Verilog language has two forms of the procedural assignment statement:

1. Blocking
2. Nonblocking

The two are distinguished by the = and <= assignment operators. The blocking assignment statement (= operator) acts much like in traditional programming languages. The whole statement is done before control passes on to the next statement. The nonblocking (<= operator) evaluates all the right-hand sides for the current time unit and assigns the left-hand sides at the end of the time unit.

Execution of blocking assignments can be viewed as a one-step process:

Evaluate the RHS (right-hand side equation) and update the LHS (left-hand side expression) of the blocking assignment without interruption from any other Verilog statement.

A blocking assignment “blocks” trailing assignment in the same block and always block from occurring until after the current assignment has been completed

Execution of nonblocking assignments can be viewed as a two-step process:

1. Evaluate the RHS of nonblocking statements at the beginning of the time
2. Update the LHS of nonblocking statements at the end of the time step

- **Ques: 31. Write the stimulus for blocking and nonblocking assignments.**

Solution:

```
module blocking;
reg [0:7] A, B;
initial
begin: init1
A = 3;
#1 A = A + 1; // blocking procedural assignment
B = A + 1;
$display ("Blocking: A= %b B= %b", A, B); A = 3;

    #1 A <= A + 1; // non-blocking procedural assignment
B <= A + 1;
#1 $display ("Non-blocking: A= %b B= %b", A, B);
end
endmodule

    output:
Blocking: A= 00000100 B= 00000101
Non-blocking: A= 00000100 B= 00000100
```

The effect is for all the nonblocking assignments to use the old values of the variables at the beginning of the current time unit and to assign the registers new values at the end of the current time unit. This reflects how register transfers occur in some hardware systems.

The blocking procedural assignment is used for combinational logic and nonblocking procedural assignment for sequential.

- **Ques: 32. What do you understand by Verilog file "open?"**

Solution:

OPEN A FILE

```
    integer file;
file = $fopenr("filename");
file = $fopenw("filename");
file = $fopena("filename");
```

The function $fopenr opens an existing file for reading. $fopenw opens a new file for writing, and $fopena opens a new file for writing where any data will be appended to the end of the file. The file name can be either a quoted-string or a reg holding the file name. If the file was successfully opened, it returns an integer containing the file number (1..MAX_FILES) or NULL (0) if there was an error. Note that these functions are not the same as the built-in system function $fopen which opens a file for writing by $fdisplay. The files are opened in C with 'rb', 'wb', and 'ab' which allows reading and writing binary data on the PC. The 'b' is ignored in Unix.

- **Ques: 33. What do you understand by Verilog file "close?"**

Solution:

CLOSE A FILE

```
integer file, r;
r = $fcloser(file);
r = $fclosew(file);
```

The function $fcloser closes a file for input. $fclosew closes a file for output. It returns EOF if there was an error, otherwise 0. Note that these are not the same as $fclose which closes files for writing.

- **Ques: 34. What are the differences between continuous and procedural assignments?**

Solution:

S. No.	Continuous Assignment	Procedural Assignment
1	The values are assigned primarily to nets.	The values are assigned to reg variables.
2	It is used for combinational logic	It is used for combinational as well as sequential circuits like flip-flops and latches.
3	The assignment occurs whenever the value of the RHS of expression changes as a continuous process.	The value of a previous assignment is held until another assignment is made to the variable.
4	It occurs in assignments such as wire, net, and port.	It occurs in constructs such as always, initial, task, and function.
5	For example: wire y1 = x1 & x2;	For example: always@ (posedge clk) reg1 <= x1;

- **Ques: 34. What are the differences between initial and always?**

Solution:
While both ***initial*** and ***always*** constructs are procedural assignments, they differ in the following ways:

initial	always
Assignments in an initial block begin to execute from lime 0 in simulation, and proceed in the specified sequence.	Assignments in an always block also begin from time 0, and repeat forever as a function of the changes on the blocks sensitivity list
Execution of statements in an initial begin-end block slops when the end of the block is reached, i.e., executed only once during simulation	Execution continuously repeats from the begin to the end of the process unless held by a ***wait*** construct throughout the simulation session
Non-synthesizable construct	Synthesi/able construct

For example,
`reg [1:0] out1, out2;`
`initial begin`
`  out1 = 2'b10;`
`  #5 out2 = 2'b01;`
`end`
	For example,
	`reg [1:0] out1, out2;`
	`always @(posedge elk)`
	`begin`
	`  out1 <= in1;`
	`  out2 < out1 & in2;`
	`end`

- **Ques: 35. What is the difference between task and function?**

Solution:
Both tasks and functions in Verilog help in executing common procedures from different places in a module. They help in writing cleaner and maintainable code, by avoiding replication at different places in a module. Essentially, functions and tasks provide a "subroutine" mechanism of reusing the same section of code at different places in a module. This allows for easier maintenance of the code.

S.No	Task	Function
1	Task contains time control statements like @(posedge), delay operator (#)	It executes in zero simulation tune.
2	It can call functions or tasks within itself.	It can call functions within itself.
3	It cannot return any value when called; instead the task can have output arguments.	It returns a single value when called in system Verilog the return value can be optionally voided.
4	For example, vc_result is an output of a task to calculate the result of the greater of two input arguments g1 and g2 greater_val (g1, g2, vc_result);	For example, vc_result is assigned the return of a function call to calculate the result of the greater of two input arguments g1 and g2 vc_result = greater_val (g1. g2);

- **Ques: 36. Are tasks and functions re-entrant, and how are they different from static task and function calls? Illustrate with an example.**

Solution:
In Verilog-95, tasks and functions were not re-entrant. From Verilog version 2001 onward, the tasks and functions are re-entrant. The re-entrant tasks have a keyword automatic between the keyword task and the name of the task. The presence of the keyword automatically replicates and allocates the variables within a task dynamically for each task entry during concurrent task calls, i.e., the values don't get overwritten for each task call. Without the keyword, the variables are allocated statically, which means these variables are shared across different task calls, and can hence get overwritten by each task call.

The following example illustrates the effect of the keyword automatic for re-entrant tasks. This is a nonsynthesizable code for illustration only.

```
// automatic variables
  my_value = in_value; // blocking assignment
#5
  $display("my_value = \t%0d, t = %0d",
             my_value, $time);
  out_value = my_value + 2;
end
endtask

initial begin
  fork
    begin // First parallel call
      #1
      $display("in1= \t\t%0d, t = %0d", 2, $time);
      modify_value(2, out_val);
    end
    begin // Second parallel call
      #2
      $display("in2= \t\t%0d, t = %0d", 3, $time);
      modify_value(3, out_val);
    end
  join
end

endmodule
```

In the above example, my_value is a local variable in the task modify_value. Whenever this task is called, the input in_value is assigned to the local variable after 5 simulation time units. Within the initial begin, there is a fork-join, which launches two parallel processes. One starts after simulation time unit #1, and other after #2. The first process assigns a value of 2 to the output of the task, and the second one assigns a value of 3 to the output of the task. The display sequence will be:

```
in1=          2, t=1 // passed value is 2
in2=          3, t=2
my_value =    3, t=6 // retained value is 3
my_value =    3, t=7
```

- **Ques: 37. What is the sequence of events without keyword "automatic?"**

Solution:
The sequence of events without keyword automatic is as follows:

1. The launch of the two processes from the fork-join happens from time 0.
2. The first process calls modify_value after #1, and the local variable my_value is assigned the value 2. This happens at t = 1.
3. The second process calls modify_value after #2 and the local variable my_value is assigned the value 3. This happens at t = 2. Note that the earlier value of 2 assigned to the local variable my_value is now overwritten with the value 3.
4. After 4 more time units i.e., at t = 1 + 5 = 6, the display of the first task call becomes active. Since the latest value is now "3," based on the previous step, the value of "3" is displayed for my_value, instead of what was passed as "2."
5. Similarly, for the second process i.e., 2 + 5 = 7, the display of the second task call becomes active. Since the latest value is still "3," the value of "3" is displayed for my_value here too.

The critical replacement happened in step 3 above, wherein the launch of the process overwrote the value of the first process before its turn to display. This occurred because, without the automatic keyword, the variables within the task were static, and shared by all calls to the task.

Now, with the keyword automatic between the task and task name, the following is the output:

```
in1=          2, t=1 // passed value is 2
in2=          3, t=2
my_value =    2, t=6 // passed value 2 preserved
my_value =    3, t=7
```

Following the same steps as above, this time, due to the presence of the keyword automatic, the unique values of the variables are preserved in each call, and not overwritten by the subsequent task calls before the variable is being used.

- **Ques: 37. What is the difference between re-entrant and static task?**

Solution:

Reentrant task	**Static task**
Has the keyword ***automatic*** between the ***task*** keyword and identifier	*Doesn't* have the keyword ***automatic*** between the ***task*** keyword and the identifier
Variables declared within the task are allocated dynamically for each concurrent task call	Variable declarations within the task are allocated statically
All variables will be replicated in each concurrent call to store state specific to that invocation	Each concurrent call to the task will OVERWRITE the statically allocated local variables of the task from all other concurrent calls to the task
Variables declared are de-allocated at the end of task invocation	Variables retain their values between invocations
Task items cannot be accessed by hierarchical inferences	Task items can be accessed by hierarchical inferences
Task items shall be allocated new across all uses of the task executing concurrently	Task items can be shared across all uses of the task executing concurrently

- **Ques: 38. What are the rules governing usage of a Verilog *function*?**

Solution:

The following rules govern the usage of a Verilog **function** construct:

A function cannot advance simulation-time, using constructs such as #, @. etc.

A function shall not have nonblocking assignments.

A function without a range defaults to a one-bit reg for the return value.

It is illegal to declare another object with the same name as the function in the scope where the function is declared.

- **Ques: 39. What do you mean by "defparam?"**

Solution:

The defparam statement modifies the parameter value only at compilation time. The constant expression is the new value for the parameter and can contain a previously declared parameter.

In this method, the parameter within a module is accessed by its hierarchical name from anywhere within the scope of the hierarchy. In the following example, the lower level module parameter_list gets instantiated in the example_defparam module. But the values of width and depth are overridden using the defparam construct.

Advantages of using the defparam approach:

1. The ordered sequence need not be maintained in overriding the parameter values.
2. A specific parameter can be overridden rather than respecifying all the parameters before the one that's being overridden.
3. It can help with code maintenance by grouping all the defparam's collectively in a single place, which can be compiled with the rest of the code.

Parameter redefinition at instantiation is the recommended style by most expert Verilog users.

- **Ques: 40. Why "defparam" is avoided for parameter redefinition?**

Solution:
There are several reasons to avoid using defparam for parameter redefinition. Some of the reasons are listed hereunder:

1. The defparam statements if not collectively present in one place can be buried in any module, anywhere in the design hierarchy, making code difficult to maintain or reuse (a form of spaghetti code, which should always be avoided).
2. Since the defparam statements can be buried anywhere in the hierarchy, they can prevent the Verilog language compilers from being able to do a true independent compilation of the modules.
3. Since multiple defparam statements can be made to the same parameter instance, the final value of the parameter in this situation can (and probably will be) different with different tools.
4. The defparam statements are not supported in the official IEEE 1364.1-2002 synthesis subset for Verilog.
5. The IEEE 1364 standards committee is considering a proposal to deprecate defparam in the next version of the Verilog standard, making the defparam an obsolete construct.

- **Ques: 41. What are the differences between using 'define, and using either parameter or defparam for specifying variables?**

Solution:

Both ‘define and parameter constructs can be used to specify constants in the design. For example, the width parameter can be specified either as a ‘define or parameter, as:

```
'define width 64
if ('width == 64)  ...
or
parameter width 64;
if (width == 64)  ...
```

However, the following are a few differences in using the two constructs:

‘define	*parameter/defparam*
‘define is basically a text substitution macro	***Parameter*** is used to specify constants in a design
Multiple ***‘defines*** to the same variable name are not allowed, the final value of the macro is determined by source code order	Although mulliple/wrawj ftr definitions to the same variable are not allowed within a module, multiple ***defparam's*** to the same variable are allowed, however the final value of the parameter is indeterminate
Cannot be overridden in any mechanism	***Parameter*** can be overridden
Only one constant with the given name can exisl in the full scope	Multiple modules can have the same ***parameter*** name, as it is limited to lhat scope only

- **Ques: 42. What is the difference between inter statement and intra statement delay?**

Solution:

```
//define register variables
reg a, b, c;
//intra assignment delays
initial
begin
a = 0; c = 0;
b = #5 a + c; //Take value of a and c at the time=0, evaluate
//a + c and then wait 5 time units to assign value
//to b.
end
```

```
//Equivalent method with temporary variables and regular delay control
initial
begin
a = 0; c = 0;
temp_ac = a + c;
#5 b = temp_ac; //Take value of a + c at the current time and
//store it in a temporary variable. Even though a and c
//might change between 0 and 5,
//the value assigned to b at time 5 is unaffected.
end
```

- **Ques: 43. Difference between $monitor, $display, and $strobe?**

Solution:
These commands have the same syntax and display text on the screen during simulation. $display and $strobe display once every time they are executed, whereas $monitor displays every time one of its parameters changes.

The difference between $display and $strobe is that $strobe displays the parameters at the very end of the current simulation time unit rather than exactly where it is executed. The format string is like that in C/C++ and may contain format characters. Format characters include %d (decimal), %h (hexadecimal), %b (binary), %c (character), %s (string) and %t (time), %m (hierarchy level). %5d, %5b, etc., would give exactly 5 spaces for the number instead of the space needed. Append b, h, o to the task name to change the default format to binary, octal, or hexadecimal.

Syntax:

```
$display ("format_string", par_1, par_2, ... );
$strobe ("format_string", par_1, par_2, ... );
$monitor ("format_string", par_1, par_2, ... );
```

- **Ques: 44. What is the difference between full case and parallel case?**

Solution:
A "full" case statement is a case statement in which all possible case-expression binary patterns can be matched to a case item or a case default. If a case statement does not include a case default and if it is possible to find a binary case expression that does not match any of the defined case items, the case statement is not "full."

A "parallel" case statement is a case statement in which it is only possible to match a case expression to one and only one case item. If it is possible to find a case expression that would match more than one case item, the matching case items are called "overlapping" case items and the case statement is not "parallel."

full_case	**parallel_case**
Indicates that the case statements has been fully specified, and all unspecified case expressions can be optimized away	Indicates that all case items need to be evaluated in parallel and not infer any priority encoding logic
All control paths are specified explicitly or by using a default	There is no overlap among the case items
Helps avoid latches as all cases are fully specified	Results in multiplexor logic as a parallel logic
Although not recommended, the default clause can be avoided, and still not infer a latch	A priority encoder is NOT synthesized, as each path is unique
An example of a case statement that is full (and parallel) is shown below:	An example of a case statement that is parallel (and full) is shown as follows:

`reg var1 [1:0];`
`always @ (a or b or c)`
`begin`
`  case (var1)`
`   2'b00 : out1 = a;`
`   2'b01 : out1 = b;`
`   2'b10 : out1 = c;`
`   2'b11 : out1 = a&b;`
`   endcase`
`end`
	`reg var1 [2:0];`
	`always @ (a or b or c)`
	`begin`
	`  case (var1)`
	`   3'b000 : out1 = a;`
	`   3'b001 : out1 = b;`
	`   3'b010 : out1 = c;`
	`// rest of the cases are`
	`// not defined`
	`  endcase`
	`end`

Note that the ***default*** clause was not required here as it is fully specified (although having it is a good coding practice).	Note that the above case **doesn't** have a ***default*** clause; but each branch is definitely **unique**, but all cases are not specified, that is, branches missing for 2,3,4,5,6,7. The out1 register will get synthesized into a latch.

- **Ques: 45. What is meant by inferring latches, how to avoid it?**

Solution:
Consider the following:
always @(s1 or s0 or i0 or i1 or i2 or i3)
case ({s1, s0})
2'd0 : out = i0;
2'd1 : out = i1;
2'd2 : out = i2;
endcase

In a case statement, if all the possible combinations are not compared and the default is also not specified like in the example above a latch will be inferred, a latch is inferred because to reproduce the previous value when an unknown branch is specified.

For example, in the above case if {s1, s0}=3, the previously stored value is reproduced for this storing a latch is inferred.

The same maybe observed in the IF statement in case an ELSE IF is not specified.

To avoid inferring latches make sure that all the cases are mentioned if not default condition is provided.

- **Ques: 46. What do you mean by the sensitivity list?**

Solution:
The sensitivity list indicates that when a change occurs to any one of the elements in the list change, begin…end statement inside that always block will get executed.

- **Ques: 47. What logic is inferred when multiple assign statements are targeting the same wire?**

Solution:
It is illegal to specify multiple ***assign*** statements to the same ***wire*** in a synthesizable code that will become an output port of the module. The synthesis tools give a syntax error that a net is being driven by more than one source. For example, the following is illegal:

```
wire tmp;
assign tmp = in1 & in2; // only one type of
                              // output assignment is
assign tmp = in1 | in2; // legal for synthesis
```

However, it is legal to drive a three-state wire by multiple ***assign*** statements, as shown in the following example:

```
input enable1, enable2;
wire tmp;
assign tmp = (enable1 == 1'b1) ?
             (in1 & in2) : 1'bz;
assign tmp = (enable2 == 1'b1) ?
             (in3 | in4) : 1'bz;
```

- **Ques: 48. What do conditional assignments get inferred into?**

Solution:
Conditionals in a continuous assignment are specified through the "?:" operator. Conditionals get inferred into a multiplexor. For example, the following is the code for a simple multiplexer:

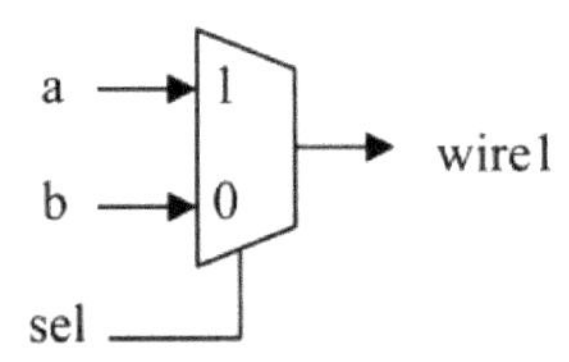


```
wire wire1;
assign wire1 = (sel == 1'b1) ? a : b;
```

- **Ques: 49. What is the logic that gets synthesized when conditional Operators in a single continuous assignment are nested?**

Solution:
Conditional operators in a single continuous assignment can be nested as shown in the following example. The logic gets elaborated into a tree of multiplexers.

```
input sel1, sel2, sel3, in1, in2, in3, in4;
output out1;
assign out1 = (sel1, == 1'b1) ? in1 :
              (sel2, == 1'b1) ? in2 :
              (sel3, == 1'b1) ? in3 : in4;
```

In the multiplexor units shown, it follows the logic that when sel is high, the output Z selects A, else selects B.

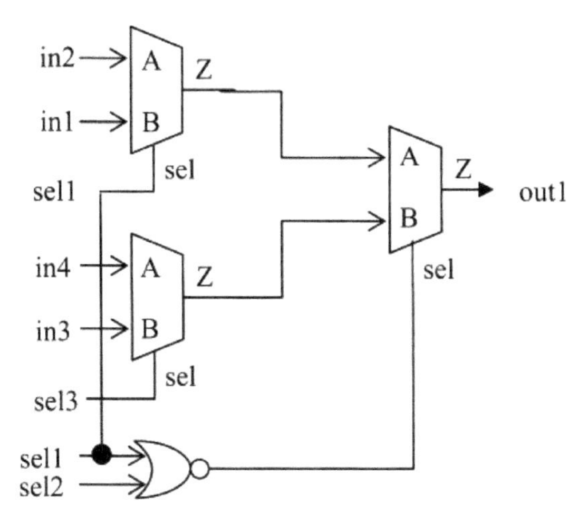

- **Ques: 50. Give the examples of a flip-flop with synchronous and asynchronous set and reset.**

Solution:

- Simple D flip-flop

Positive edge triggered, no set or reset, the value of Q is unknown at power on

```
module dff (clk, d, q);
input clk, d;
output q;

reg q;
always @(posedge clk) begin
  q <= d;
end
endmodule
```

In SystemVerilog, the same code would be implemented with ***always_ff*** in place of the ***always*** keyword, as follows:

```
always_ff @(posedge clk) begin
  q <= d;
end
```

The advantage of ***always_ff*** over ***always*** is that ***always_ff*** indicates that the designer intends to model clocked sequential logic. Software tools can then verify that the block's sensitivity list and functionality correctly represent the type of logic intended.

2. Asynchronous set FF

Positive edge triggered, active high asynchronous set

```
module asff (clk, d, set, q);
input clk, d, set;
output q;

reg q;
always @(posedge clk or posedge set) begin
  if (set)
    q <= 1'b1;
  else
    q <= d;
end
endmodule
```

Replacing the ***always*** keyword with ***always_ff*** above would implement the asynchronous FF in SystemVerilog.

3. Asynchronous reset FF

Positive edge triggered, active high asynchronous reset

```
module arff (clk, d, reset, q);
input clk, d, reset;
output q;

reg q;
always @(posedge clk or posedge reset) begin
  if (reset)
    q <= 1'b0;
  else
    q <= d;
end
endmodule
```

Replacing the ***always*** keyword with ***always_ff*** above would implement the asynchronous FF in SystemVerilog.

4. Asynchronous set and reset FF

Positive edge triggered, active high asynchronous set and reset

```
module arsff(clk, d, set, reset);
input clk, d, set, reset;
output q;
```

```
reg q;
always @(posedge clk or posedge set or
         posedge reset)
begin
  if (set)
    q <= 1'b1;
  else if (reset)
    q <= 1'b0;
  else
    q <= d;
end
endmodule
```

Replacing the ***always*** keyword with ***always_ff*** above would implement the asynchronous FF in SystemVerilog.

5. Synchronous set FF

   Positive edge triggered, active high synchronous set

```
module ssff (clk, d, set, q);
input clk, d, set;
output q;

reg q;
always @(posedge clk) begin
  if (set)
    q <= 1'b1;
  else
    q <= d;
end
endmodule
```

Replacing the ***always*** keyword with ***always_ff*** above would implement the asynchronous FF in SystemVerilog.

6. Synchronous reset FF

   Positive edge triggered, active high synchronous reset

```
module srff (clk, d, reset, q);
input clk, d, reset;
output q;

reg q;
always @(posedge clk) begin
  if (reset)
    q <= 1'b0;
```

```
  else
    q <= d;
end
endmodule
```

Replacing the ***always*** keyword with ***always_ff*** above would implement the asynchronous FF in SystemVerilog.

7. Synchronous set and reset FF

Positive edge triggered, active high synchronous set and reset

```
module ssrff (lk, d, set, reset, q);
input clk, d, set, reset;
output q;

reg q;
always @(posedge clk) begin
  if (set)
    q <= 1'b1;
  else
    q <= d;
end
endmodule
```

Replacing the ***always*** keyword with ***always_ff*** above would implement the asynchronous FF in SystemVerilog.

- **Ques: 51. List the various keywords, compiler directive, and system task available in Verilog.**

Solution:

List of Keywords, System Tasks, and Compiler Directives Used in Verilog HDL

This list consists of keywords, system task, and compiler directives. All the keywords are defined in lowercase. The system tasks are *tasks* and functions that are used to generate input and output during the simulation. The compiler directives are used to control the compilation of a *Verilog* description. The reference is IEEE std. 1364-2001, Verilog HDL.

Keywords		System Task	Compiler Directives
always	module	$bitstoreal	'accelerate
assign	task	$countdrivers	'autoexpand_vectornets
begin	library	$display	'celldefine
fork	time	$fclose	'default_nettype

Keywords		System Task	Compiler Directives
case	table	$fdisplay	’define
casex	specify	$fmonitor	’define
casez	join	$fopen	’else
buf	end	$fstrobe	’elsif
bufif0	endcase	$fwrite	’endcelldefine
bufif1	endtable	$finish	’endif
rtran	endprimitive	$getpattern	’endprotect
rtranif0	endmodule	$history	’endprotected
rtranif1	endspecify	$incsave	’expand_vectornets
defparam	endtask	$input	’ifdef
deassign	pull0	$itor	’ifndef
include	pull1	$key	’include
integer	pullup	$list	’noaccelerate
instance	pulldown	$log	’noexpand_vectornets
automatic	tri	$monitor	’noremove_gatenames
cell	tri0	$monitoroff	’nounconnected_drive
cmos	tri1	$monitoron	’protect
pmos	force	$nokey	’protected
nmos	forever	$stop	’remove_gatenames
and	real	$finish	’remove_netnames
or	reg	$write	’resetall
not	repeat	$rtoi	’timescale
nand	if	$readmemh	’unconnected_drive
nor	else	$readmemb	’undef
strong0	parameter	$hold	
strong1	primitive	$period	
supply0	wait	$skew	
supply1	wire	$timeformat	

- **Ques: 52. What is the difference between latch and flip-flop?**

Solution:

Latch	**Flip-flop**
Area of a latch is typically less lhan that of a Flip-flop	Area of a Flip-Hop for same features is more than that of a lalch
Consumes lesser power, due to lesser switching activity and lesser area	Power consumption is typically higher, due lo the area and free running clock. Additional controls required to save power
Facilitates time borrowing or cycle stealing; Helps increase pipeline depth with lesser area.; Even if the path is longer than a clock cycle for a latch based pipeline, it is okay as long as il meets the next latch setup margin	Since the clock boundaries arc rigid, the facility of time borrowing or cycle stealing doesn't exist with FFs. A negative slack cannot be propagated to the liming of the next stage in pipeline and hence must execute within a clock period
In multiple clock schemes, the clock edges must not be overlapping; It makes the logic design, vector generation for verification and clock tree synthesis difficult	Clock tree synthesis is less tedious in FF based designs. Since the stimulus needs lo be stable before the setup time of the clock, the vector generation is relatively easier
With lime borrowing* and cycle stealing, the operating frequency is higher than the slowest logic path	Due to rigid timing boundaries, the slowest path pretty much decides the operating frequency
Makes time budgeting arid characterizing the interfaces tedious	The time budgeting is clearer and characterizing the interface is easier

- **Ques: 53. How do I choose between a case statement and a multiway if-else statement?**

Solution:

Both case and if-else are flow control constructs. Functionally in the simulation, they yield similar results. While both these constructs get elaborated into combinatorial logic, the usage scenarios for these constructs are different.

A case statement is typically chosen for the following scenarios:

1. When the conditionals are mutually exclusive and only one variable controls the flow in the case statement. The case variable itself could be a concatenation of different signals.
2. To specify the various state transitions of a finite state machine

3. Use of casex and casez allows the use of x and z to represent don't-care bits in the control expression

A multiway if the statement is typically chosen in the following scenarios:

1. Synthesizing priority encoded logic
2. When the conditionals are not mutually exclusive and more general in using multiple expressions for the condition expression.

The advantages of using the case over if-else are as follows:

1. Case statements are more readable than if-else
2. When used for state machines, there is a direct mapping between the state machine's "bubble diagram" and the case description.

In a case construct, if all the possible cases are not specified, and the default clause is missing, a latch is inferred. Likewise, for an if-else construct, if a final else clause is missing, a latch is inferred.

- **Ques: 54. What is the difference in using (== or !=) vs. (===or !==) in the decision making of a flow control construct in a synthesizable code?**

Solution:
In Verilog, the (==) operator is called logical equality, and (!=) is called logical inequality operator. The (===) operator is called case equality, and (!==) is called case inequality. The following are the differences in using these constructs in synthesizable code.

Use of = or != operators	Use of == or != operators
These operators can be used in a synthesi/able code	Cannot be used in a synthesizable code
If cither of the operands have x or z value, the result is unknown	The operands will be compared, even if they have x and z values in the bits
If any of the operators is x or z, the logical result of comparison is **always** FALSE	The x and z bits will be used in comparison, and the logical result will be a TRUE or FALSE, based on actual comparison
Since the operands contain x and z, the result will he an x. Hence, the comparison can be non-deterministic	Since x and z are also used in comparison, the result of comparison will be Boolean 1 or 0. Hence the comparison can be deterministic

Use of = or != operators	Use of == or != operators

Example of using (== or !=) operators
`if (a == b)`
`  out1 = a & b;`
`else`
`  out1 = a | b;`
	Example of using (=== or !==) operators
	`if (a === b)`
	`  out1 = a & b;`
	`else`
	`  out1 = a | b;`

If either a or b becomes x or z, the else clause will be executed and `out1` will be driven by OR gate	If a and b are identical, even if they becomes x or z, the if clause will be executed and `out1` will be driven by AND gate

- **Ques: 55. Explain the differences and advantages of casex and casez over the case statement?**

Solution:
The ***casex*** operator has to be used when both the high impedance value (z) and unknown (x) in any bit has to be treated as a don't-care during case comparisons. The ***casez*** operator treats the (z) operator as a don't-care during case comparisons.

In both cases, the bits that are treated as don't-care will not be considered for comparison, that is, only bit values other than don't care bits are used in the comparison. The wildcard character "?" can be used in place of "z" for literal numbers.

The following is an example of a ***casex*** statement

```
input [2:0] in1;
reg [2:0] reg1;
casex (in1)
  3'b0x0 : out1 = a & b; // same as conditions
                         // 3'b010, 3'b000
  3'bx10 : out1 = a | b; // same as conditions
                         // 3'b110, 3'b010
  default : out1 = a ^ b // for all other
                         // conditions
endcase
```

The same example, if written with an if-else tree, would look like:

```
// bit in1[1] is not considered at all
  if (!in1[2] & !in1[0]) out1 = (a & b);
// bit in1[2] is not considered
  else if (in1[1] & !in1[0]) out1 = (a | b);
// default clause
else out1 = (a ^ b);
```

Using ***casex*** or ***casez*** has the following coding advantages:

1. It reduces the number of lines, especially if the number of bits had been more
2. Makes code look more clear and less cluttered
3. It simplifies the optimization, as it is clear that the bits with x are to be ignored.

- **Ques: 56. What are the differences between synchronous and asynchronous state machines?**

Solution:

Asynchronous state machines	Synchronous slate machines
State transitions depend upon the order in which the input signals change	State transitions are controlled by a clock signal
State transitions happen after propagation delay of the state line	State transitions happen at intervals of the clock period
Delay lines act as memory elements	Edge triggered FFs or level sensitive latches act as storage elements
Output response time is not predictable	Output response lime Is predictable; will happen at clock period intervals

- **Ques: 57. Illustrate the differences between Mealy and Moore state machines.**

Solution:
Both the Mealy machine and Moore machine are two commonly used coding styles of state machines. The basic block diagram of these two state machines are shown as follows:

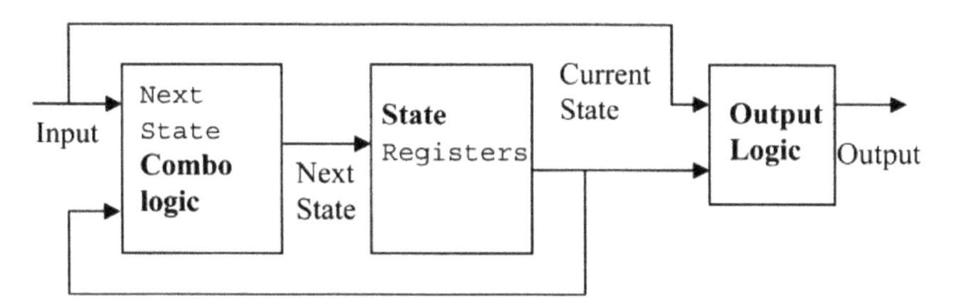

Block diagram of a Mealy machine.

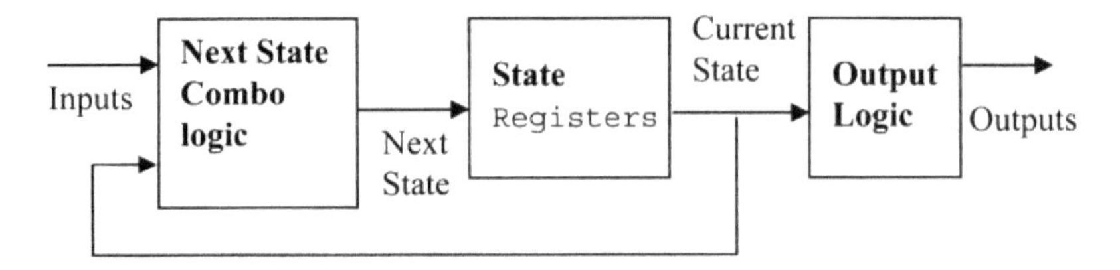

Block diagram of a Moore machine.

- **Ques: 58. What is the difference between the Mealy machine and the Moore machine?**

Solution:

Mealy machine	Moore machine
Outputs are a function of current state and input signals	Outputs are a function of current state only
Output can change between changes between state	Outputs change only when the current state changes
Output can changes any number of times during a clock cycle, which may result in glitches on the outputs	Output is delayed by one clock cycle, but is stable
More output combinations are possible as the outputs are a function of inputs too	Since the outputs are a function only of the current state, the numbers of output combinations arc fewer with the Mealy machine

- **Ques: 59. Illustrate the differences between binary encoding and one hot encoding mechanism state machines.**

Solution:
The encoding in state machines is primarily either binary [sometimes called sequential] encoding or the one-hot encoding. Both mechanisms eventually lead to decoding of the states, but their logic implementation, timing and area implications differ. The differences are summarized in the table as follows:

Binary encoding	One-hot encoding
Requires fewer number of FFs to represent current slate	Number of FFs required is equal to the number of states in the FSM
As there is combinatorial palh in the output logic, its timing is not as good as the one-hot encoding mechanism	Better output timing, as there is no output logic. Only clk→q delay, and hence faster

Binary encoding	One-hot encoding
Preferred approach in ASICs unless the timing in output path is critical	Useful and necessary in register rich application like FPGAs
Since the number of FFs is limited, good optimization is required for encoding	Don't need to optimize the state encoding, as each state has unique flop anyway.
Adding or deleting stales requires tracking the side effects to the other states in the FSM	Easy to maintain, that is, adding or deleting stales is easy, and doesn't effect the rest of the states
Tedious to debug, since a wrong state transition needs a walk through of the next state combinatorial logic	Easy to debug, since a wrong slate transition can be easily detected by looking at ihe current slate values
Critical path analysis requires tracking the combinatorial logic	Easy to find critical paths during Static Timing Analysis (STA)

- **Ques: 60. What is the difference between synchronous memories and Asynchronous memories?**

Solution:

Synchronous memories	Asynchronous memories
Data writes and reads based on a clock port	Data writes and reads typically based on an enable pin
Has better static timing, because the data output from a synchronous memory is registered	The data output from an asynchronous memory is a combinatorial lookup of its address inputs; Therefore, this combinatorial logic could potentially become a critical element in the liming path
Has larger area compared to asynchronous memory	Area is less compared to synchronous memory
Read operations are generally two clock cycles, minimum; the first cycle is usually used by the memory to sample the address, and the second cycle will be used by the external system to sample the read-data	Both read and write cycles are asynchronous, based on "enable" pins

- **Ques: 61. What are a few power reduction techniques that can be implemented during the backend analysis?**

Solution:

The following are a few parameters within the chip that can significantly influence the overall power consumption, which can be taken care of during the backend phase:

Have shorter routes for power and timing-sensitive logic: Since the capacitance of a routing net is a function of the length, width, and impedance of the route, a long route typically has higher capacitance than the shorter alternative. Since dynamic power consumed is directly proportional to the capacitance, that is, $P = CV^2f$, lower capacitance means lesser power. This would mean the logic blocks need to be closer to each other.

Reduce excessive loading: Heavily loaded nets cause higher capacitance and higher power consumption.

- **Ques: 62. What are a few power reduction techniques that can be implemented during board design?**

Solution:

The following are a few techniques that can reduce power consumption at a board level:

1. Reduce the chip interconnection dynamic power by limiting the number of I/O pins, the loading on each, pin and the average frequency at which each pin toggles.
2. Minimize the trace lengths between the chips output and other device inputs.

- **Ques: 63. What is the difference between assign-deassign and force-release?**

Solution:

The assign-deassign and force-release constructs in Verilog have similar effects but differ in the fact that force-release can apply to nets and variables, whereas assign-deassign applies only to variables.

The procedural assign-deassign construct is intended to be used for modeling hardware behavior, but the construct is not synthesizable by most logic synthesis tools. The force-release construct is intended for design verification and is not synthesizable.

- **Ques: 64. Does Verilog support an (a^b) operator?**

Solution:

Yes. Verilog supports the a^b operation by using two astrices, back to back. This operator was added with the Verilog-2001 release. For example,

```
module powerof (in1, out1);
parameter power = 2;
input [1 : 0] in1;
output [3 : 0] out1;

assign out1 = in1 ** power;

endmodule // powerof
```

A value of 2 would mean out1 = in1 * in1, that is, the value getting multiplied to itself. Simulation, however, works for powers other than 2, as well.

- **Ques: 65. What is the difference between $strobe and $monitor?**

Solution:
The differences between $strobe and $monitor are summarized in the following points:

- $strobe can be used to create new simulation events, simply by encapsulating the $strobe system call within a simulation construct that moves simulation time, such as @(posedge clock), @(negedge clock),@(any_signal), etc. There can exist multiple $strobe system calls at the same time, with identical or different arguments.
- $monitor stands alone. A given set of arguments of $monitor form their unique sensitivity list. Only one $monitor call can be active at any time. Each call to $monitor replaces any previous call(s) to $monitor.

- **Ques: 66. What is PLI? Why it is used?**

Solution:
Programming Language Interface (PLI) of Verilog HDL is a mechanism to interface Verilog programs with programs written in C language. It also provides a mechanism to access internal databases of the simulator from the C program.

PLI is used for implementing system calls which would have been hard to do otherwise (or impossible) using Verilog syntax. Or, in other words, you can take advantage of both the paradigms – parallel and hardware related features of Verilog and sequential flow of C – using PLI.

- **Ques: 67. Differentiate between VHDL and Verilog.**

Solution:
Verilog and VHDL are HDLs. They both can be used to represent the behavioral of a digital circuit.

Verilog:

It is a loosely typed language. Mean data types can be used in a mix and match way.

It is case-sensitive language. “A” and “a” are not the same in Verilog.

The syntax is similar to that of “C” language.

There is no library management is Verilog.

There are fewer constructs in Verilog.

Serves better for low-level modeling.

VHDL:

It is a strongly typed language. Opposite to that of Verilog.

It is case insensitive.

There are a lot more constructs in VHDL as compared to Verilog.

Serves best for high-level modeling.

VHDL is popular with (European) FPGA designers because low-level modeling is not required in an FPGA flow. ASIC designers prefer Verilog.

- **Ques: 68. Differentiate between C and HDL with the help of graphical representation.**

Solution:

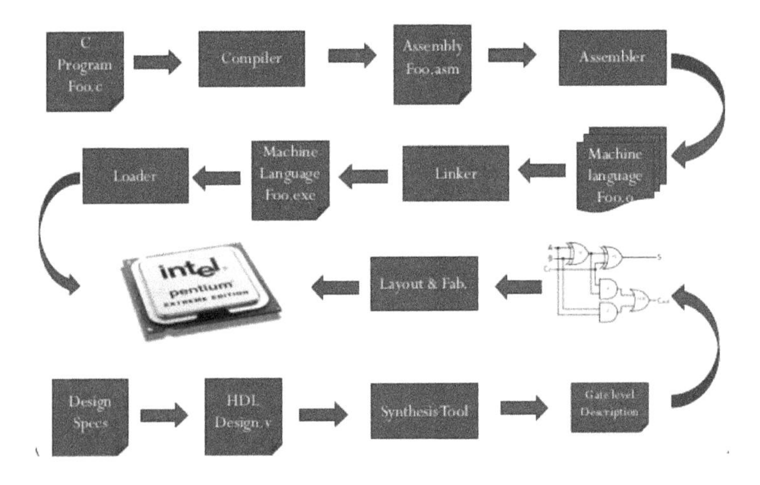

In the C program, there is a series of instructions loaded into memory and to be executed by the microprocessor in sequential order.

In the C program, there is no provision of timing control to control its operation because the next statement will not be executed until finishing the current statement execution.

Verilog HDL will configure hardware to do a certain function. Verilog statements may have parallel concurrency or sequential execution or both. Verilog code must be synthesizable, i.e., compiler must be able to generate the logic that fits the description.

Verilog HDL has timing control as gate delays are there. Statements maybe executed parallelly.

- **Ques: 69. State the main objectives of HDL.**

Solution:

1. Allow for testing/verification using computer simulation.
   It includes syntax for timing and delay.
2. Allow for synthesizing.
   Synthesizable HDL

- **Ques: 70. How many primary data types in Verilog HDL?**

Solution:

1. Nets: Represents structural connections between the components.
2. Registers: It represents a variable for data storage.

- **Ques: 71. Explain the data types associated with signals.**

Solution:

1. Explicitly declared: with the declaration in Verilog code
2. Implicitly declared: It is always net-type "wire" and one bit wide. It has no declaration when used to connect structural building blocking in code.

- **Ques: 72. Why should we use a nonblocking assignment for sequential instead of blocking assignments?**

Solution:

To overcome race around the condition. Let us explain the situation with an example:

```
reg R1, R2, R3, R4;       always @(posedge clk)
always @(posedge clk)     R2 <= R1;
R2 = R1;                  always @(posedge clk)
always @(posedge clk)     R3 <= R2;
R3 = R2                   always @(posedge clk)
always @(posedge clk)     R4 <= R3;
R4 = R3
```

BLOCKING **NON-BLOCKING**

In blocking assignment, code run in some order but the sequence is not known. But in the case of a nonblocking assignment, RHS is evaluated when assignment runs. LHS updated only after all events for the current instant have run.

- **Ques: 73. Differentiate between inter assignment delay and intra assignment delay.**

Solution:
Inter assignment delay:
Inter assignment are those delay statements where the execution of the entire statement or assignment got delayed. Inter assignment delays in Verilog often corresponds to the inertial delay or the regular delay statements in VHDL.

```
#5 x = y + z;
```

Here in the above statement, its entire execution got delayed by 5-time units.

Intra assignment delay:

```
x = #5 y + z;
```

In such delays, it calculates the values of RHS expression at the current time unit and delays the assignment of the calculation to the LHS. Here in the example, the addition of y and z is calculated at the 0-time step and then it waits for 5-time units. After that, it assigned the corresponding value of the expression to x. Intra statement delay corresponds to the transport delay of VHDL.

- **Ques: 74. Generate clock in Verilog.**

Solution:
Clocks are the main synchronizing events to which all other signals are referenced. The clock can be generated in many ways.

Example 1:

```
initial clk = 0;
always #10 clk = ~clk;
```

Example 2:

```
always
begin
clk = 0;
#10 clk = 1;
end
```

- **Ques: 75. Explain the significance of Verilog in VLSI.**

Solution:
Verilog is an HDL for describing electronic circuits and systems. In Verilog, circuit components are prepared inside a module. It contains both behavioral and structural statements. Structural statements signify circuit components like logic gates, counters, and microprocessors. Behavioral statements represent programming aspects like loops, if-then statements, and stimulus vectors.

- **Ques: 76. In Verilog code what does "timescale 1 ns/ 1 ps" signify?**

Solution:
In Verilog code, the unit of time is 1 ns and the accuracy/precision will be up to 1ps.

- **Ques: 77. Mention the two types of procedural blocks in Verilog.**

Solution:
The two types of procedural blocks in Verilog are
Initial: Initial blocks runs only once at time zero
Always: This block loop to execute over and again and executes always, as the name suggests.

- **Ques: 78. Explain why present VLSI circuits use MOSFETs instead of BJTs.**

Solution:
In comparison to BJT, MOSFETS can be made very compact as they occupy a very small silicon area on IC chip and also in terms of manufacturing, they are relatively simple. Moreover, digital and memory ICs can be employed with circuits that use only MOSFETs, i.e., diodes, resistors, etc.

- **Ques: 79. Mention what are three regions of operation of MOSFET and how are they used.**

Solution:
MOSFET has three regions of operations

1. Cut-off region
2. Triode region
3. Saturation region

The triode and cut-off region are used to function as a switch, while saturation region is used to operate as an amplifier.

- **Ques: 80. Explain why is the number of gate inputs to CMOS gates usually limited to four.**

Solution:
The higher the number of stacks, the slower the gate will be. In NOR and NAND gates the number of gates present in the stack is usually alike as the number of inputs plus one. So, input is restricted to four.

- **Ques: 81. Explain what is slack.**

Solution:
Slack is referred to as a time delay difference from the expected delay to the actual delay in a particular path. Slack can be negative or positive.

- **Ques: 82. Explain the use of defpararm.**

Solution:
With the keyword defparam, parameter values can be configured in any module instance in the design.

- **Ques: 83. How many abstraction levels are there in VLSI?**

Solution:
Every system should be decomposed into *three* fundamental domains:

1. Behavioural domain
2. Structural domain
3. Physical domain

- **Ques: 84. Explain the level of abstraction using Onion diagram/ Y chart.**

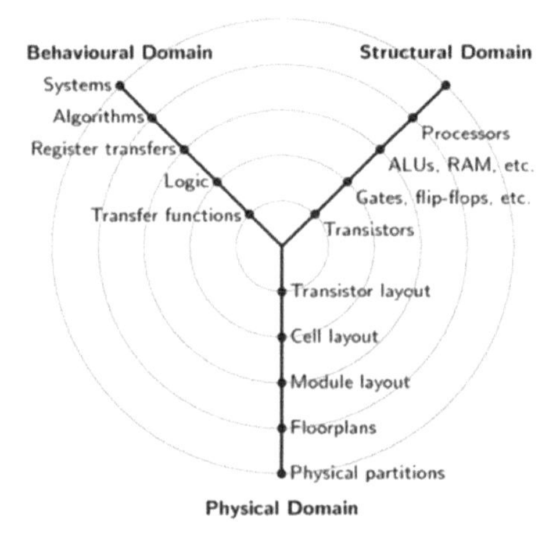

Gajski-Kuhn Y-chart

Solution:
Gajski Kuhn Y chart is also known as the Onion diagram.

We can design the system at various layers, which are called design abstraction levels:

1. Architecture
2. Algorithm
3. Modules (or functions)
4. Logic
5. Switch
6. Circuit
7. Device

- **Ques: 85. Explain simulation and synthesis.**

Solution:
Simulation is to verify the functionality of the circuit.
Synthesis is to convert HDL description into a set of primitives (equation in PLD/component in FPGA) to fit in the target technology. Synthesis means the construction of a gate-level netlist from the model of the circuit, described in HDL. It converts behavior into the structure.

References

1. Mano, M. M. and Ciletti, M. (2013). *Digital Design: With an Introduction to the Verilog HDL*. Pearson.
2. Palnitkar, S. (2003). *Verilog HDL: A Guide to Digital Design and Synthesis* (Vol. 1). Prentice-Hall Professional.
3. Ciletti, M. D. (2003). *Advanced Digital Design with the Verilog HDL* (Vol. 1). Upper Saddle River: Prentice-Hall.
4. Bhasker, J. (1999). *A Verilog HDL Primer*. Star Galaxy Publishing.
5. Chonnad, S. S. and Balachander, N. B. (2007). *Verilog: Frequently Asked Questions: Language, Applications and Extensions*. Springer Science & Business Media.

5

Miscellaneous Section

Electronics and electrical engineering is an emerging and most popular engineering discipline. Electronics engineering and electrical engineering are closely related to this field. Even imagining a life bereft of electronic gadgets seems impossible in today's world. There is no field left across the globe where one cannot find the usage of electronics and communication engineering. Perhaps that is why electronics have become the vertebrae of digital technology.

5.1 Branches of Electronics

Electronics is further having emerging branches which target various industry and interest.

5.1.1 Electronics and Communication Engineering

This branch deals with analog and digital transmission and reception of data, networking, voice and video, solid-state devices, microprocessors, digital and analog communication, satellite communication, antennae, and wave progression. It also deals with the manufacturing of electronic devices, circuits, and communications equipment such as transmitter, receiver, integrated circuits (ICs), microwaves, and fiber among others.

An electronics engineer is qualified for jobs that include building electronic components for integration into larger systems, IC design work, board layout, programming microcontrollers and microprocessors, and field testing.

Electronics and communication engineers deal with all of the applications which make our life easier and enjoyable such as television, radio, computers, mobiles, etc., are designed and developed by electronics and communication engineers.

5.1.2 Electronics and Telecommunication Engineering

This branch deals with various principles and practical aspects related to designing various telecommunication equipment. Electronics and telecommunication engineers develop prototypes of IC components. Apart from this, it also combines various aspects of electrical, structural, and civil engineering as well.

Telecommunications engineers handle different types of technology that help us to communicate. This includes research, design and develop satellite and cable systems, mobile phones, radio waves, the internet, and e-mail.

Electronics and telecommunication engineers deal with consumer electronics, aviation and avionics, manufacturing, electricity generation and distribution, communications, transportation, and telecommunications industry as electronics engineers.

5.1.3 Microelectronics and VLSI Design

Microelectronics is a field of specialization and VLSI is a process under this specialization. VLSI design deals with "design for VLSI," it could be analog, digital, mixed-signal, etc, which is again a process under microelectronics. Microelectronics tends to focus on a circuit that is developed for a specific purpose, while VLSI tends to focus on multiple ICs to perform a more difficult or more general purpose. (very large systems integration).

An example of a microelectronic would be an accelerometer that uses electronic tunneling to output a voltage based on acceleration. An example of a VLSI would be the microcontroller that is used to interpret and utilize the accelerometer signal.

Technical Questions with Solutions

- **Ques: 1. What are the advantages of VLSI design?**

Solution:
The number of transistors in an IC has dramatically increased.

VLSI integrates all of these into one chip resulting in the following advantages:

1. Reduces the size of circuits
2. Reduces the cost of the devices
3. Increases operating speed of circuits

- **Ques: 2. What is the future of VLSI technology?**

Solution:
The future of VLSI circuits depends on the trend of channel length reduction. Available fabrication technologies deny more degradation in channel length; so, nanoelectronics devices such as QCA (quantum-dot cellular automata, SET (single electron transistor), CNTFET (carbon nanotube field-effect transistors), and benzene rings are candidates for replacement of conventional CMOS technology.

- **Ques: 3. What is the transistor count in the latest VLSI technology?**

Solution:
The transistor count is the number of transistors on an IC. The rate at which transistor counts have increased generally follows Moore's law, which observed that the transistor count doubles approximately every two years.

In the latest 7 nm process technology, the APPLE has released Apple A12X Bionic Processor (Octacore AMD 64 mobile SOC) which has 10,000,000,000 transistors in 122m^2 area.

- **Ques: 4. What is the leakage current in CMOS?**

Solution:
The leakage current is an unwanted conductive path under normal operating conditions. When Vgs<Vth, the current leaks between drain and source of MOS. The acceptance value of leakage current is 210 μA.

There are four main leakage currents in a CMOS transistor:

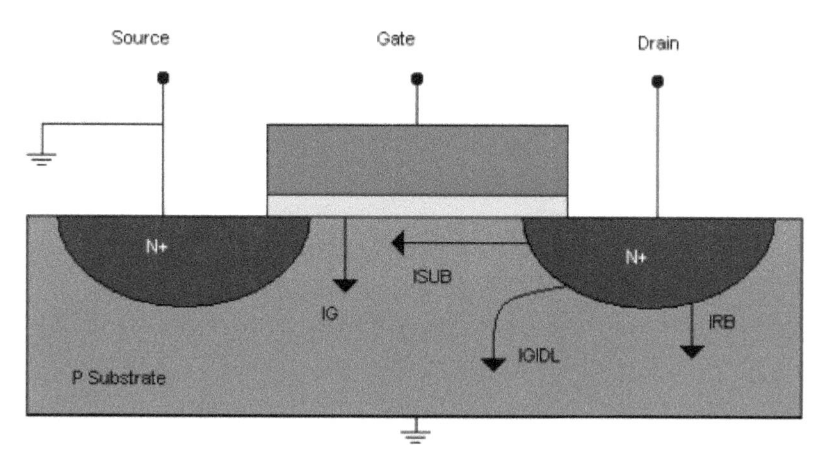

A. Subthreshold (weak inversion) leakage (ISUB)

The subthreshold leakage current is a drain to source current of a transistor when it operates in the weak inversion region. It is due to the diffusion current of minority carriers in the channel of a MOS structure. When the gate voltage is below Vth, the turn-on or threshold voltage, an NMOS device is turned off.

B. Reverse-biased Source/Drain junction leakage (IRB)

Recalling the structure of an NMOS transistor, a reverse-biased p–n junction exists that is formed by either the source or drain to the substrate when it is off. Although a potential barrier exists, there is a leakage current. It has two components: electron–hole generation in the depletion region and minority carrier diffusion/drift at the edge of the same depletion region.

C. Gate direct-tunneling leakage

The current due to gate direct-tunneling leakage is current tunneling into the gate of the transistor. The mechanisms for this tunneling effect are electron conduction band tunneling (ECB), electron valence band tunneling (EVB), and hole valence band tunneling (HVB). Currently, ECB is the dominant phenomenon. There is electron tunneling because of the high electric fields that exist across the oxide layer.

D. Gate-induced drain leakage

The current due to gate-induced drain leakage (GIDL) is a direct consequence of the high field effect in the drain of a MOS transistor. For example, an NMOS with its gate grounded and drain at the supply voltage potential experiences energy band bending in the drain. This allows electron–hole pairs to be generated by avalanche effects and band-to-band carrier tunneling. Holes are quickly driven to the body, creating a deep depletion situation. Similarly, electrons are collected in the drain. Together, these effects create the IGIDL current.

- **Ques: 5. What do you mean by parasitic capacitance in MOSFET?**

Solution:
These are unwanted capacitances, but still are part of the transistor. Together with the resistances in the circuit, they put an upper limit to the speed of the transistor.

When two electrical conductors at different voltage are close together, the electric field between them causes electric charge to be stored on them. This is known as the parasitic capacitance effect.

Types of parasitic capacitances

1. Junctions depletion capacitances
2. Overlap and gate-channel capacitances
3. Channel-bulk depletion capacitance

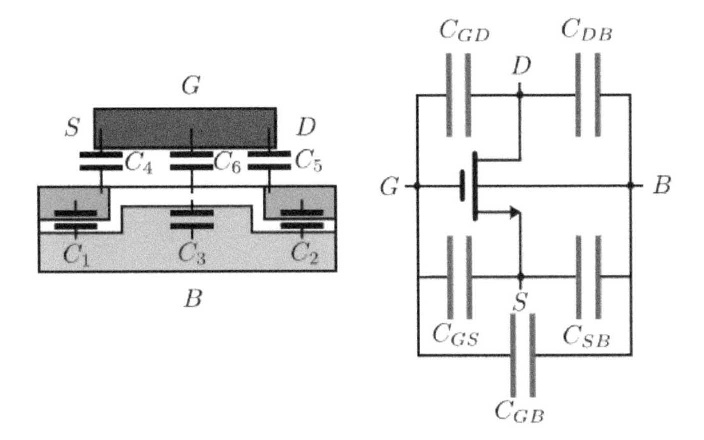

C1 and C2 are capacitances created by the depletion regions between source/drain and bulk. C3 is the depletion capacitance between the channel and bulk. C4 and C5 are capacitances caused by the overlap between the gate and the source/drain diffusions. Finally, C6 is the oxide capacitance between the gate and the channel and is split between drain and source depending on the region of operation of the transistor.

- **Ques: 6. What do you mean by FINFET?**

Solution:
FinFET, also known as fin field effect transistor, is a type of nonplanar or "3D" transistor used in the design of modern processors. As in earlier, planar designs, it is built on an SOI (silicon on insulator) substrate. However, FinFET designs also use a conducting channel that rises above the level of the insulator, creating a thin silicon structure, shaped like a fin, which is called a gate electrode. This fin-shaped electrode allows multiple gates to operate on a single transistor. Intel began releasing FinFET CPU technology in 2012 with its 22-nm Ivy Bridge processors.

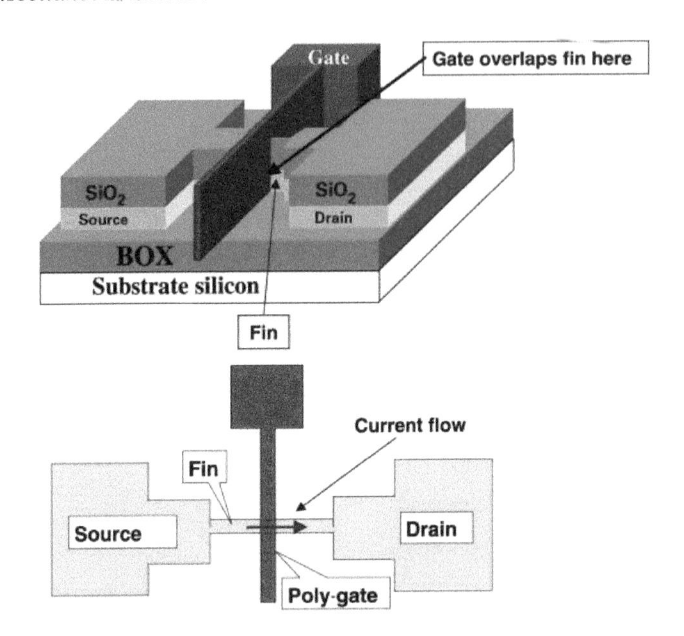

Dr. Chenming Hu has been called the father of 3D transistor for developing the FinFET in 1999.

- **Ques: 7. What are the benefit and drawbacks of FINFET?**

Solution:

☐ To exploit different benefits of FinFET, it is fabricated into two types:

(1) Dual-gate FinFET, which trims the excess silicon by fabricating the channel using an ultrathin layer of silicon that sits on top of an insulator, therefore the electric field from the gate to the fin on the top is drastically reduced.

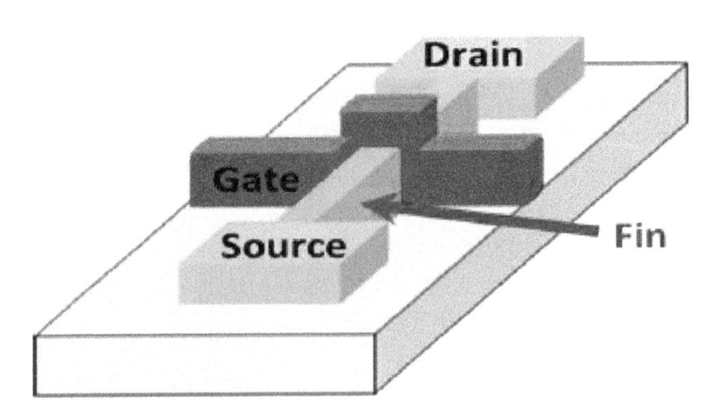

(2) Tri-gate FinFET, in which the FET gate wraps around three sides of the transistor's elevated channel or "fin." Since fins are made vertical, high packing density can be achieved, by packing transistors closer together. Further, to get even more performance and energy-efficiency gains, designers also can continue growing the height of the fins.

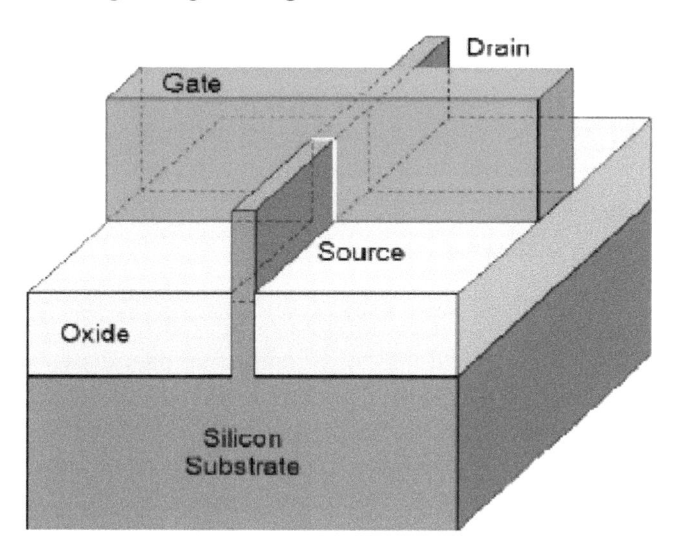

- ☐ It exhibits little or no body effect because FinFET channels are fully depleted.
- ☐ Given the excellent control of the conducting channel by the gate, very little current is allowed to leak through the body when the device is in the off state. The FinFET can also be run at a lower operating voltage for given leakage current, halving its dynamic power consumption (which is proportional to CV2f) for a 0.7 scaling in VDD.
- ☐ At 1 V, the FinFET is 18% faster than the equivalent planar device, but at 0.7 V, the advantage is 37%. This enables the device to be operated at lower threshold voltages for the same leakage.
- ☐ The difference between the gate and threshold voltage at very low operating voltages is much greater, thus exaggerating the performance advantage of very low-voltage FinFETs.
- ☐ Increased voltage headroom for circuits such as cascades
- ☐ Lower gate resistance, which helps keep flicker noise under control, as well as improved matching
- ☐ Higher current drive and higher gain

However, disadvantages are the following:

- The designer cannot control the channel as easily and the higher source/drain resistance cuts transconductance.
- Designers have little choice overvoltages for I/O and have to develop more complex methods to achieve ESD immunity.
- The Si surface of fins appears different than in bulk, therefore excessive Si loss was observed after the usual pregate-oxide clean. Thus, wet cleans are optimized with dilute concentration and lower temperatures. Similarly, the oxidation of fins is also faster at the corner and tip of fins.
- Besides, the dry etching on fins is more stringent due to the 3D structures and a bias plasma pulsing scheme may be viable for minimizing Si loss.

- **Ques:8. What is skin effect?**

Solution:
Every conductor line has resistance and inductance! When you pass AC in a conductor (imagine a rod), then due to its alternating nature and with the conductor inductance, it finds some opposition for the flow of current in the regions closer to the center of the conductor. Since this inductive reactance is stronger at the center, the current tends to avoid that area and prefers flowing on the periphery or the skin of the conductor. Hence, skin effect. The more the frequency, the stronger is the reactance.

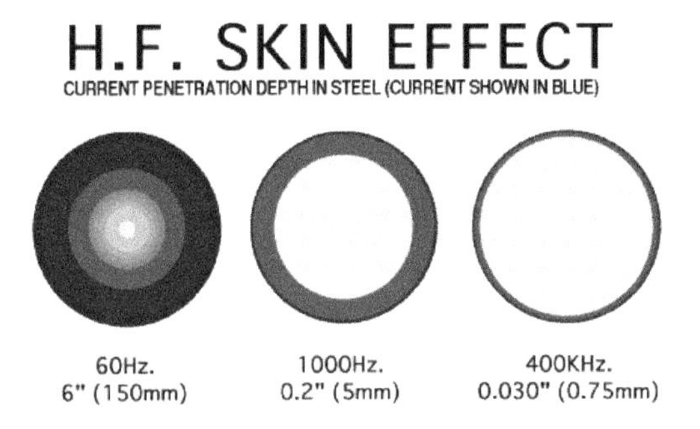

- **Ques: 9. Why skin effect occurs in AC, but not in DC?**

Solution:
The skin effect is caused by the back emf produced by the self-induced magnetic flux in a conductor. For a DC, the rate of change of flux is zero, so there is no back emf due to changes in magnetic flux. Therefore, the current is uniformly distributed throughout the cross-section of the conductor.

In the case of DC, there is no change in frequency and hence the inductive reactance is zero. So, the current doesn't find any opposition other than the resistance and it flows completely in the conductor, not just outside.

- **Ques: 10. What do you mean by MuGFET?**

Solution:
A multigate device or multiple-gate field-effect transistor (MuGFET) refers to a MOSFET (metal–oxide–semiconductor field-effect transistor) that incorporates more than one gate into a single device. The multiple gates maybe controlled by a single gate electrode, wherein the multiple-gate surfaces act electrically as a single gate, or by independent gate electrodes. A multigate device employing independent gate electrodes is sometimes called a multiple-independent-gate field-effect transistor (MIGFET).

- **Ques: 11. What is a tri-gate transistor or 3D transistor?**

Solution:
Tri-gate or 3D transistor (not to be confused with 3D microchips) fabrication is used by Intel Corporation for the nonplanar transistor architecture used in Ivy Bridge, Haswell, and Skylake processors. These transistors employ a single gate stacked on top of two vertical gates (a single gate wrapped over 3 sides of the channel), allowing essentially three times the surface area for electrons to travel. Intel reports that their tri-gate transistors reduce leakage and consume far less power than current transistors. This allows up to 37% higher speed or power consumption at under 50% of the previous type of transistors used by Intel.

- **Ques: 12. What is electronics? Define it.**

Solution:
Electronics is the branch of science that deals with the study of flow and control of electrons (electricity) and the study of their behavior and effects in vacuums, gases, and semiconductors, and with devices using such electrons. This control of electrons is accomplished by devices that resist, carry, select, steer, switch, store, manipulate, and exploit the electron.

- **Ques: 13. What is communication?**

Solution:
Communication means transferring a signal from the transmitter which passes through a medium then the output is obtained at the receiver or communication is transferring a message from one place to another place.

- **Ques: 14. Explain types of communication.**

Solution:
There are two types of communication: analog and digital communication.

As a technology, analog is the process of taking an audio or video signal (the human voice) and translating it into electronic pulses. Digital, on the other hand, is breaking the signal into a binary format where the audio or video data is represented by a series of "1"s and "0s."

Digital signals are immune to noise, quality of transmission and reception is good, as components used in digital communication can be produced with high precision and power consumption is also very less when compared with analog signals.

- **Ques: 15. What are active and passive components? Give some examples.**

Solution:
Passive: Capable of operating without an external power source. Typical passive components are resistors, capacitors, inductors, and diodes (although the latter is a special case).

Active: Requiring a source of power to operate. It includes transistors (all types), ICs (all types), TRIACs, SCRs, light-emitting diodes (LEDs), etc.

- **Ques: 16. What are DC and AC?**

Solution:
DC: Direct current. The electrons flow in one direction only. Current flow is from negative to positive, although it is often more convenient to think of it as from positive to negative. This is sometimes referred to as "conventional" current as opposed to electron flow.

AC: Alternating current. The electrons cyclically flow in both directions – first one way, then the other. The rate of change of direction determines the frequency measured in Hertz (cycles per second).

- **Ques: 17. Define the term "frequency?"**

Solution:
Frequency: Unit is Hertz, symbol is Hz, the old symbol was cps (cycles per second). A complete cycle is completed when the AC signal has gone from zero volts to one extreme, back through zero volts to the opposite extreme, and returned to zero. The accepted audio range is from 20 Hz to 20,000 Hz. The number of times the signal completes a complete cycle in 1 sec is the frequency.

- **Ques: 18. What do you mean by "voltage?"**

Solution:
Voltage: Unit is Volts, symbol is V or U, the old symbol was E . Voltage is the "pressure" of electricity, or "electromotive force" (hence the old term E). A 9 V battery has a voltage of 9 V DC and maybe positive or negative depending on the terminal that is used as the reference. The mains have a voltage of 220, 240, or 110 V depending where you live – this is AC, and alternates between positive and negative values. Voltage is also commonly measured in millivolts (mV), and 1,000 mV is 1 V. Microvolts (UV) and nanovolts (nV) are also used.

- **Ques: 19. What do you mean by "current?"**

Solution:
Current: Unit is Amperes (Amps), symbol is I. Current is the flow of electricity (electrons). No current flows between the terminals of a battery or other voltage supply unless a load is connected. The magnitude of the current is determined by the available voltage and the resistance (or impedance) of the load and the power source. Current can be AC or DC, positive or negative, depending upon the reference. For electronics, current may also be measured in mA (milliamps) – 1,000 mA is 1 A. Nanoamps (nA) is also used in some cases.

- **Ques: 20. What do you mean by "resistance?"**

Solution:
Resistance: Unit is Ohms, symbol is R or Ω. Resistance is a measure of how easily (or with what difficulty) electrons will flow through the device. Copper wire has a very low resistance, so a small voltage will allow a large current to flow. Likewise, the plastic insulation has a very high resistance and prevents current from flowing from one wire to those adjacent. Resistors have a defined resistance, so the current can be calculated for any voltage. Resistance in passive devices is always positive (i.e., >0).

- **Ques: 21. Give the SI units and symbols of various electronic components.**

Solution:

Name	Symbol	Quantity	In other SI units	In SI base units
radian	rad	plane angle	1	$(m{\cdot}m^{-1})$
steradian	sr	solid angle	1	$(m^2{\cdot}m^{-2})$
hertz	Hz	frequency		s^{-1}

Name	Symbol	Quantity	In other SI units	In SI base units
newton	N	force, weight		$kg{\cdot}m{\cdot}s^{-2}$
pascal	Pa	pressure, stress	N/m^2	$kg{\cdot}m^{-1}{\cdot}s^{-2}$
joule	J	energy, work, heat	$N{\cdot}m = Pa{\cdot}m^3$	$kg{\cdot}m^2{\cdot}s^{-2}$
watt	W	power, radiant flux	J/s	$kg{\cdot}m^2{\cdot}s^{-3}$
coulomb	C	electric charge or quantity of electricity		$s{\cdot}A$
volt	V	voltage (electrical potential), emf	W/A	$kg{\cdot}m^2{\cdot}s^{-3}{\cdot}A^{-1}$
farad	F	capacitance	C/V	$kg^{-1}{\cdot}m^{-2}{\cdot}s^4{\cdot}A^2$
ohm	Ω	resistance, impedance, reactance	V/A	$kg{\cdot}m^2{\cdot}s^{-3}{\cdot}A^{-2}$
Siemens	S	electrical conductance	Ω^{-1}	$kg^{-1}{\cdot}m^{-2}{\cdot}s^3{\cdot}A^2$
weber	Wb	magnetic flux	$V{\cdot}s$	$kg{\cdot}m^2{\cdot}s^{-2}{\cdot}A^{-1}$
tesla	T	magnetic flux density	Wb/m^2	$kg{\cdot}s^{-2}{\cdot}A^{-1}$
henry	H	inductance	Wb/A	$kg{\cdot}m^2{\cdot}s^{-2}{\cdot}A^{-2}$
degree Celsius	°C	temperature relative to 273.15 K		K
lumen	Im	luminous flux	$cd{\cdot}sr$	cd
lux	Ix	illuminance	lm/m^2	$m^{-2}{\cdot}cd$

- **Ques: 22. Define 1 ohm.**

Solution:
1 ohm is the unit of resistance. It is equal to 1 volt upon 1 ampere. It means that an object having a resistance of 1ohm allows 1 ampere of current to flow through 2 points having a potential difference of 1 volt.

- **Ques: 23. Define 1 ampere.**

Solution:
1 Ampere: One ampere is defined as the current that flows with an electric charge of one Coulomb per second. or, the ampere is that current which, when passing through a resistance of 1 ohm, produces a potential difference of 1 V across its terminals.

- **Ques: 24. Define 1 watt.**

Solution:
If a circuit is having pure resistance and in which we apply a 1-volt voltage through which 1-ampere current is produced by load then power consumption is 1 watt.

Power is defined as the rate of doing work.

$$Power = \frac{work}{time}$$

1 watt is the power of an appliance that consumes energy at the rate of 1 joule per second.

- **Ques: 25. Define 1 Joule.**

Solution:
Joule is an SI unit of work. 1 Joule is the amount of work done when a force of 1 Newton displaces a body through a distance of 1 m in the direction of the force applied.

- **Ques: 26. Define 1 volt.**

Solution:
One volt is defined as energy consumption of one joule per electric charge of one coulomb. 1 V = 1 J/C. One volt is equal to the current of 1 amp times the resistance of 1 ohm: 1 V = 1 A · 1 Ω

One volt is the difference in electrical potential between two points of conducting wire when an electric current of 1 A dissipates 1 W power between those points.

- **Ques: 27. Define 1 farad.**

Solution:
One farad is defined as the capacitance across which, when charged with one coulomb, there is a potential difference of one volt.

$$C = \frac{Q}{V}$$

The 1-farad capacitor can store one columb of charge at 1 volt.

- **Ques: 28. Define 1 columb.**

Solution:
It is the basic unit of electric charge, equal to the quantity of charge transferred in one second by a steady current of one ampere, and equivalent to 6.2415 × 1018 elementary charges, where one elementary charge is the charge of a proton or the negative of the charge of an electron.

Amount of charge transported by a constant current of 1 Amp in 1 second is 1 columb.

$$Q = I \times T$$

- **Ques: 29. What is an electronic circuit?**

Solution:
A circuit is a structure that directs and controls electric currents, presumably to perform some useful function. The very name "circuit" implies that the structure is closed, something like a loop.

- **Ques: 30. Define "pressure" and "force?"**

Solution:
The pressure is an expression of force exerted on a surface per unit area. The standard unit of pressure is pascal.

$$F = M \times A \quad \text{and} \quad P = \frac{F}{A}$$

A force is any interaction that, when unopposed, will change the motion of an object.

- **Ques: 31. Define 1 Pascal and 1 Newton.**

Solution:
Pascal: Pascal is the force over an area and used as a measurement of pressure.

So, 1 Pascal = 1 Newton force spread on 1 square meter area.1 Pascal = 0.00015 PSI

Newton : Newton is the amount of force which helps in accelerating 1 kg mass by 1 meter / second. 1 Newton = 0.225 pounds.

- **Ques: 32. Define "electric charge?"**

Solution:
Electric charge is the physical property of matter that causes it to experience a force when placed in an electromagnetic field. There are two types of electric charges: positive and negative (commonly carried by protons and electrons, respectively).

- **Ques: 33. Define "power?"**

Solution:
Electrical power is the rate at which electrical energy is converted to another form, such as motion, heat, or an electromagnetic field.

- **Ques: 34. What do you understand by transducers?**

Solution:
A transducer is a device that converts energy from one form to another. Usually, a transducer converts a signal in one form of energy to a signal in another.

Transducers that convert physical quantities into mechanical ones are called mechanical transducers. Transducers that convert physical quantities into electrical are called electrical transducers.

The examples are a thermocouple that changes temperature differences into a small voltage, or a linear variable differential transformer (LVDT) used to measure displacement.

- **Ques: 35. Define "sensor" and "actuator?"**

Solution:
Sensors and actuators are comprehensive classes of transducers. Some transducers can operate as a sensor or as an actuator, but not as both simultaneously.

A sensor is a transducer that receives and responds to a signal or stimulus from a physical system. It produces a signal, which represents information about the system, which is used by some type of telemetry, information, or control system.

Sensor examples: Hotwire anemometers (measure flow velocity), microphones (measure fluid pressure), accelerometers (measure the acceleration of a structure), gas sensors (measure concentration of specific gas or gases), humidity sensor, temperature sensors, etc.

An actuator is a device that is responsible for moving or controlling a mechanism or system. It is controlled by a signal from a control system or manual control. It is operated by a source of energy, which can be a mechanical force, electrical current, hydraulic fluid pressure, or pneumatic pressure, and converts that energy into motion. An actuator is a mechanism by which a control system acts upon an environment.

Actuator examples: Motors (which impose a torque), force heads (which impose a force), pumps (which impose either a pressure or a fluid velocity).

- **Ques: 36. What is the difference between the active sensor and a passive sensor?**

Solution:
Active sensors require an external power source to operate, which is called an excitation signal. The signal is modulated by the sensor to produce an output signal. For example, a thermistor does not generate an electrical signal, but by passing an electric current through it, its resistance can be measured by detecting variations in the current or voltage across the thermistor.

Passive sensors, in contrast, generate an electric current in response to an external stimulus which serves as the output signal without the need for an additional energy source. Such examples are a photodiode, a piezoelectric sensor, thermocouple, etc.

- **Ques: 37. What do you mean by hysteresis?**

Solution:
The magnetization of ferromagnetic substances due to a varying magnetic field lags behind the field. This effect is called hysteresis, and the term is used to describe any system in whose response depends not only on its current state but also upon its history.

- **Ques: 38. What is the antenna effect?**

Solution:
Increasing net length can accumulate more changes while manufacturing of the device due to the ionization process. If this net is connected to the gate of the MOSFET it can damage dielectric property of the gate and causing damage to MOSFET.

- **Ques: 39. What is cloning and buffering?**

Solution:
Cloning: It is a method of optimization that decreases the load of the heavily loaded cell by replacing the cell.

Buffering: It is a method of optimization that is used to insert buffer in high fan-out nets to decrease the delay.

- **Ques: 40. Why NAND gate is preferred over NOR?**

Solution:
At transistor level, the mobility of electrons is normally three times that of holes compared to NOR and the NAND gate is faster with less leakage.

- **Ques: 41. What is the difference between "electronics" and "electrical?"**

Solution:
Electronics deals with the flow of charge (electron) through nonmetal conductors (semiconductors).

Electrical deals with the flow of charge through metal conductors.

Example: Flow of charge through silicon which is not a metal would come under electronics, whereas the flow of charge through copper which is a metal would come under electrical.

- **Ques: 42. What do you mean by resistor, capacitor, and inductor?**

Solution:
Resistors: A resistor is an electrical device that resists the flow of electrical current. It is a passive device used to control, or impede the flow of, electric current in an electric circuit by providing resistance, thereby developing a drop in voltage across the device. The value of a resistor is measured in ohms and represented by the Greek letter capital omega.

Capacitors: In simple words, we can say that a capacitor is a device used to store and release electricity, usually as the result of chemical action. Also referred to as a storage cell, a secondary cell, a condenser, or an accumulator. A Leyden jar was an early example of a capacitor.

Inductors: An inductor is an electrical device (typically a conducting coil) that introduces inductance into a circuit. An inductor is a passive electrical component designed to provide inductance in a circuit. It is a coil of wire wrapped around an iron core. The simplest form of an inductor is made up of a coil of wire. The inductance measured in henry is proportional to the number of turns of wire, the wire loop diameter, and the material or core the wire is wound around. It stores energy in the form of a magnetic field.

- **Ques: 43. What do you mean by semiconductor devices?**

Solution:
A conductor made with semiconducting material. Semiconductors are made up of a substance with electrical properties intermediate between a good conductor and a good insulator. A semiconductor device conducts electricity poorly at room temperature but has increasing conductivity at higher temperatures. Metalloids are usually good semiconductors.

- **Ques: 44. Why silicon is preferred over germanium?**

Solution:
The knee voltage of germanium is 0.7 eV and silicon is 1.1 eV because the atomic size of germanium is bigger. Lesser energy is required to free electrons from the outermost orbit.

1. Low Reverse Leakage Current

The reverse current in silicon flows in order of nano amperes compared to germanium in which the reverse current is in order of microamperes, because of this the accuracy of nonconduction of the Ge diode in reverse bias falls. In contrast, Si diode retains its property to a greater extent, i.e., it allows the negligible amount of current to flow.

2. Good Temperature Stability

Temperature stability of silicon is good, it can withstand in temperature range typically from 140 °C to 180 °C, whereas germanium is much temperature-sensitive; i.e., it can withstand only up to 70 °C.

3. Low Cost

Silicon is relatively easy and inexpensive to obtain and process, whereas germanium is a rare material that is typically found with copper, lead, or silver deposits. Because of its rarity, germanium is more expensive to work with, thus making germanium diodes more difficult to find (and sometimes more expensive) than silicon diodes.

4. High Reverse Break Down Voltage

The Si diode has a large reverse breakdown voltage of about 70–100 V compared to Ge, which has the reverse breakdown voltage around 50 V.

5. Large Forward Current

Silicon is much better for high current applications as it has very high forward current in a range of tens of amperes, whereas germanium diodes have very small forward current in a range of microamperes.

- **Ques: 45. What do you mean by the P–N junction diode? Draw its VI characteristics.**

Solution:
A P–N junction diode is formed when a p-type semiconductor is fused to an n-type semiconductor creating a potential barrier voltage across the diode junction.

If a suitable positive voltage (forward bias) is applied between the two ends of the P–N junction, it can supply free electrons and holes with the extra energy they require to cross the junction as the width of the depletion layer around the P–N junction is decreased.

A diode is a one-way valve for electricity. Diodes allow the flow of electricity in one direction.

Knee voltage is also known as "cut-in-voltage." The minimum amount of voltage required for conducting the diode is known as "knee voltage" or "cut-in-voltage." The forward voltage at which the current through P–N junction starts increasing rapidly is known as knee voltage.

Biasing conditions

1. Zero bias – No external voltage potential is applied to the P–N junction diode.
2. Reverse bias – The voltage potential is connected negative (–ve) to the P-type material and positive (+ve) to the N-type material across the diode which has the effect of increasing the P–N junction diode's width.
3. Forward bias – The voltage potential is connected positive (+ve) to the P-type material and negative (–ve) to the N-type material across the diode which has the effect of decreasing the P–N junction diode's width.

V–I Characteristics

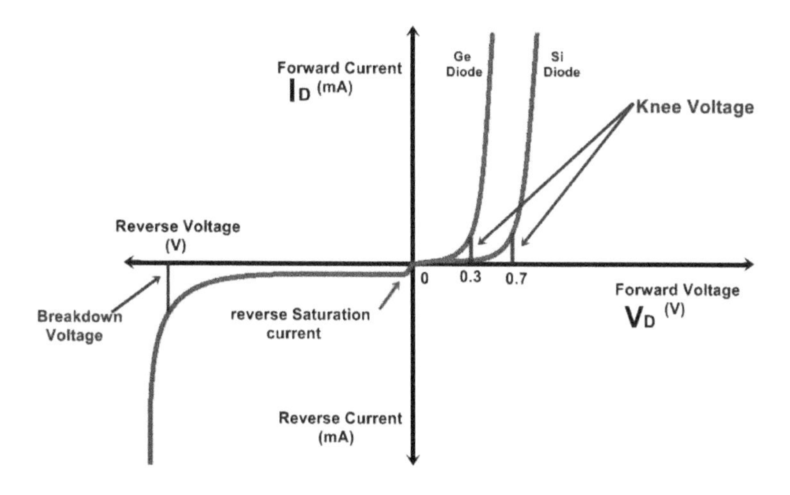

- **Ques: 47. Give the Shockley diode equation.**

Solution:

The Shockley diode equation or the diode law, named after transistor coinventor William Shockley of Bell Telephone Laboratories, gives the I–V (current–voltage) characteristic of an idealized diode in either forward or reverse bias (applied voltage):

$$I = I_S\left(e^{\frac{V_D}{nV_T}} - 1\right)$$

where

I is the diode current,
IS is the reverse bias saturation current (or scale current),
VD is the voltage across the diode,
V_T is the thermal voltage kT/q (Boltzmann constant times temperature divided by electron charge), and *n* is the ideality factor, also known as the quality factor or sometimes emission coefficient.

The thermal voltage V_T is approximately 25.8563 mV at 300 K (27°C; 80°F). At an arbitrary temperature, it is a known constant defined by:

$$V_T = \frac{kT}{q}$$

where k is the Boltzmann constant, T is the absolute temperature of the p–n junction, and q is the magnitude of the charge of an electron (the elementary charge).

- **Ques: 48. Draw the reverse V–I characteristics of the P–N junction diode.**

Solution:

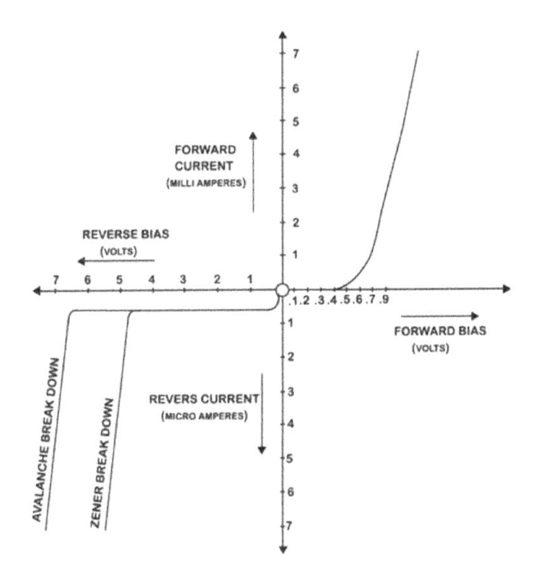

Zener Breakdown

When we increase the reverse voltage across the p–n junction diode, what happens is that the electric field across the diode junction increases (both internal and external). This results in a force of attraction on the negatively charged electrons at the junction. This force frees electrons from its covalent bond and moves those free electrons to the conduction band. When the electric field increases (with applied voltage), more and more electrons are freed from its covalent bonds. This results in drifting of electrons across the junction and electron–hole recombination occurs. So, a net current is developed and it increases rapidly with an increase in the electric field.

Zener breakdown phenomena occur in a P–N junction diode with heavy doping and thin junction (means depletion layer width is very small). Zener breakdown does not result in damage to the diode. Since the current is only due to the drifting of electrons, there is a limit to the increase in current as well.

Avalanche Breakdown

Avalanche breakdown occurs in a p–n junction diode which is moderately doped and has a thick junction (means its depletion layer width is high). Avalanche breakdown usually occurs when we apply a high reverse voltage across the diode (obviously higher than the Zener breakdown voltage, say Vz). So as we increase the applied reverse voltage, the electric field across the junction will keep increasing.

If the applied reverse voltage is Va and the depletion layer width is d, then the generated electric field can be calculated as Ea = Va/d.

This generated electric field exerts a force on the electrons at the junction and it frees them from covalent bonds. These free electrons will gain acceleration and it will start moving across the junction with high velocity. This results in a collision with other neighboring atoms. These collisions in high velocity will generate further free electrons. These electrons will start drifting and electron–hole pair recombination occurs across the junction. This results in a net current that rapidly increases.

- **Ques: 49. Why avalanche breakdown occurs at a voltage (Va) which is higher than Zener breakdown voltage (Vz)?**

Solution:
The reason behind this is simple. We know that avalanche phenomena occur in a moderately doped diode and junction width (say *d*) is high. A Zener

breakdown occurs in a diode with heavy doping and thin junction (here *d* is small). The electric field that occurs due to applied reverse voltage (say V) can be calculated as E = V/d.

So in a Zener breakdown, the electric field necessary to break electrons from the covalent bond is achieved with lesser voltage than an avalanche breakdown. The reason is a thin depletion layer width. In avalanche breakdown, the depletion layer width is higher and hence much more reverse voltage has to be applied to develop the same electric field strength (necessary enough to break electrons free)

- **Ques: 50. What do you mean by transistor?**

Solution:
The transistor is a semiconductor device. It is the fundamental building block of the circuitry in mobile phones, computers, and several other electronic devices. A transistor has a very fast response and is used in many functions including voltage regulation, amplification, switching, signal modulation, and oscillators. Transistors maybe packaged individually or they can be a part of an IC. Some of the ICs have billions of transistors in a very small area.

In electronics, a transistor is a semiconductor device commonly used to amplify or switch electronic signals. The transistor is the fundamental building block of computers and all other modern electronic devices. Some transistors are packaged individually but most are found in ICs.

- **Ques: 51. What is a printed circuit board?**

Solution:
A PCB (printed circuit board) or PC board is a piece of phenolic or glass-epoxy board with copper-clad on one or both sides. The portions of copper that aren't needed are etched off, leaving "printed" circuits that connect the components.

PCB is used to mechanically support and electrically connect electronic components using conductive pathways, or traces, etched from copper sheets laminated onto a nonconductive substrate.

It is also sometimes referred to as a printed wiring board (PWB) or etched wiring board. A PCB assembled with electronic components is called printed circuit assembly (PCA), or, printed circuit board assembly (PCBA).

- **Ques: 52. What is communication?**

Solution:
Communication means transferring a signal from the transmitter which passes through a medium and then the output is obtained at the receiver (or) communication is transferring a message from one place to another place.

- **Ques: 53. What are the different types of communications? Explain.**

Solution:
Analog and digital communication
As a technology, analog is the process of taking an audio or video signal (the human voice) and translating it into electronic pulses. Digital, on the other hand, is breaking the signal into a binary format where the audio or video data is represented by a series of "1"s and "0s."

Digital signals are immune to noise, quality of transmission, and reception is good, as components used in digital communication can be produced with high precision and power consumption is also very less when compared with analog signals.

- **Ques: 54. What is sampling?**

Solution:
The process of obtaining a set of samples from a continuous function of time x(t) is referred to as sampling.

- **Ques: 55. State sampling theorem.**

Solution:
It states that while taking the samples of a continuous signal, it has to be taken care that the sampling rate is equal to or greater than twice the cut-off frequency and the minimum sampling rate is known as the Nyquist rate.

$$Sampling\ rate \geq 2(cut - off\ frequency)$$

The minimum sampling rate is the Nyquist rate.

- **Ques: 56. What is the cut-off frequency?**

Solution:
The frequency at which the response is –3dB concerning the maximum response.

- **Ques: 57. What is passband?**

Solution:
Passband is the range of frequencies or wavelengths that can pass through a filter without being attenuated.

- **Ques: 58. What is the stopband?**

Solution:
A stopband is a band of frequencies, between specified limits, in which a circuit, such as a filter or a telephone circuit, does not let signals through, or the attenuation is above the required stopband attenuation level.

- **Ques: 59. Explain RF.**

Solution:
Radio frequency (RF) is a frequency or rate of oscillation within the range of about 3 Hz–300 GHz. This range corresponds to the frequency of alternating current electrical signals used to produce and detect radio waves. Since most of this range is beyond the vibration rate that most mechanical systems can respond to, RF usually refers to oscillations in electrical circuits or electromagnetic radiation.

- **Ques: 60. What is modulation? Where it is utilized?**

Solution:
Modulation is the process of varying some characteristic of a periodic wave with an external signal.

Radio communication superimposes this information bearing signal onto a carrier signal.

These high-frequency carrier signals can be transmitted over the air easily and are capable of traveling long distances.

The characteristics (amplitude, frequency, or phase) of the carrier signal are varied by the information-bearing signal.

Modulation is utilized to send an information-bearing signal over long distances.

- **Ques: 61. What is demodulation?**

Solution:
Demodulation is the act of removing the modulation from an analog signal to get the original baseband signal back. Demodulating is necessary because the receiver system receives a modulated signal with specific characteristics and it needs to turn it to baseband.

- **Ques: 62. Name the modulation techniques.**

Solution:
For analog modulation – AM, SSB, FM, PM, and SM Digital modulation – OOK, FSK, ASK, Psk, QAM, MSK, CPM, PPM, TCM, OFDM.

- **Ques: 63. Explain AM and FM.**

Solution:
AM: Amplitude modulation is a type of modulation where the amplitude of the carrier signal is varied following the information-bearing signal. FM: Frequency modulation is a type of modulation where the frequency of the carrier signal is varied following the information-bearing signal.

- **Ques: 64. Where do we use AM and FM?**

Solution:
AM is used for video signals, for example, TV. Ranges from 535–1705 kHz. FM is used for audio signals, for example, Radio. Ranges from 88–108 MHz.

- **Ques: 65. What is the base station?**

Solution:
The base station is a radio receiver/transmitter that serves as the hub of the local wireless network, and may also be the gateway between a wired network and the wireless network.

- **Ques: 66. How many satellites are required to cover the earth?**

Solution:
Totally 3 satellites are required to cover the entire earth, which is placed at 120 degrees to each other. The life span of the satellite is about 15 years.

- **Ques: 67. What is a repeater?**

Solution:
A repeater is an electronic device that receives a signal and retransmits it at a higher level and/or higher power, or onto the other side of obstruction so that the signal can cover longer distances without degradation.

- **Ques: 68. What is an amplifier?**

Solution:
An electronic device or electrical circuit that is used to boost (amplify) the power, voltage, or current of an applied signal.

- **Ques: 69. Give an example for negative feedback and positive feedback.**

Solution:
The example for −ve feedback is amplifiers and for +ve feedback is oscillators.

- **Ques: 70. What is an oscillator?**

Solution:
An oscillator is a circuit that creates a waveform output from a direct current input. The two main types of oscillators are harmonic and relaxation. The harmonic oscillators have smooth curved waveforms, while relaxation oscillators have waveforms with sharp changes.

- **Ques: 71. What is an IC?**

Solution:
An IC, also called a microchip, is an electronic circuit etched onto a silicon chip. Its main advantages are low cost, low power, high performance, and very small size.

- **Ques: 72. What is cross-talk?**

Solution:
Cross-talk is a form of interference caused by signals in nearby conductors. The most common example is hearing an unwanted conversation on the telephone. Cross-talk can also occur in radios, televisions, networking equipment, and even electric guitars.

- **Ques: 73. What is the resistor?**

Solution:
A resistor is a two-terminal electronic component that opposes an electric current by producing a voltage drop between its terminals in proportion to the current, that is, following Ohm's law: $V = IR$.

- **Ques: 74. What is an inductor?**

Solution:
An inductor is a passive electrical device employed in electrical circuits for its property of inductance. An inductor can take many forms.

- **Ques: 75. What is a conductor?**

Solution:
A substance, body, or device that readily conducts heat, electricity, sound, etc. Copper is a good conductor of electricity.

- **Ques: 76. What is op-amp?**

Solution:
An operational amplifier often called an op-amp, is a DC-coupled high-gain electronic voltage amplifier with differential inputs[1] and, usually, a single output. Typically the output of the op-amp is controlled either by negative feedback, which largely determines the magnitude of its output voltage gain or by positive feedback, which facilitates regenerative gain and oscillation.

- **Ques: 77. What is a feedback?**

Solution:
Feedback is a process whereby some proportion of the output signal of a system is passed (fed back) to the input. This is often used to control the dynamic behavior of the system.

- **Ques: 78. What are the advantages of negative feedback over positive feedback?**

Solution:
Much attention has been given by researchers to negative feedback process because negative feedback process lead systems toward equilibrium states. Positive feedback reinforces a given tendency of a system and can lead a system away from equilibrium states, possibly causing quite unexpected results.

- **Ques: 79. What are the Barkhausen criteria?**

Solution:
Barkhausen criteria, without which you will not know which conditions, are to be satisfied with oscillations.

"Oscillations will not be sustained if, at the oscillator frequency, the magnitude of the product of the transfer gain of the amplifier and the magnitude of the feedback factor of the feedback network (the magnitude of the loop gain) are less than unity."

The condition of unity loop gain $-A\beta = 1$ is called the Barkhausen criterion. This condition implies that

$A\beta = 1$and that the overall phase shift of feedback circuit and amplifier must be zero, i.e., phase of $-A\beta$ is zero.

It is a mathematical condition to determine whether a linear electronic circuit will oscillate or not.

- **Ques: 80. What is meant by CDMA, TDMA, and FDMA?**

Solution:
Code division multiple access (CDMA) is a channel access method utilized by various radio communication technologies. CDMA employs spread-spectrum technology and a special coding scheme (where each transmitter is assigned a code) to allow multiple users to be multiplexed over the same physical channel. By contrast, time division multiple access (TDMA) divides access by time, while frequency-division multiple access (FDMA) divides it by frequency. An analogy to the problem of multiple access is a room (channel) in which people wish to communicate with each other. To avoid confusion, people could take turns speaking (time division), speak at different pitches (frequency division), or speak in different directions (spatial division). In CDMA, they would speak different languages. People speaking the same language can understand each other, but not other people. Similarly, in radio CDMA, each group of users is given a shared code. Many codes occupy the

same channel, but only users associated with a particular code can understand each other.

- **Ques: 81. Explain different types of feedback.**

Solution:
Types of feedback:

1. Negative feedback: This tends to reduce output (but in amplifiers, stabilizes and linearizes operation). Negative feedback feeds part of a system's output, inverted, into the system's input, generally with the result that fluctuations are attenuated. Positive feedback: This tends to increase output.

2. Positive feedback, sometimes referred to as "cumulative causation," is a feedback loop system in which the system responds to perturbation (A perturbation means a system, is an alteration of function, induced by external or internal mechanisms) in the same direction as the perturbation. In contrast, a system that responds to the perturbation in the opposite direction is called a negative feedback system.

3. Bipolar feedback: This can either increase or decrease output.

- **Ques: 82. What are the main divisions of the power system?**

Solution:
The generating system, transmission system, and distribution system.

- **Ques: 83. What is the instrumentation amplifier (IA) and what are all the advantages?**

Solution:
An instrumentation amplifier is a differential op-amp circuit providing high input impedances with ease of gain adjustment by varying a single resistor.

- **Ques: 84. What is meant by the impedance diagram?**

Solution:
The equivalent circuit of all the components of the power system is drawn and they are interconnected is called impedance diagram.

- **Ques: 85. What is the need for a load flow study?**

Solution:
The load flow study of a power system is essential to decide the best operation existing system and for planning the future expansion of the system. It is also essential for designing the power system.

- **Ques: 86. What is the need for base values?**

Solution:
The components of the power system may operate at different voltage and power levels. It will be convenient for analysis of the power system if the voltage, power, current ratings of the components of the power system are expressed regarding a common value called base value.

- **Ques: 87. What is the difference between resistance and impedance?**

Solution:
The resistance exists in DC circuits since it consists of a zero frequency while impedance exists in AC circuits.

Resistance is a concept used for DC (direct currents), whereas impedance is the AC (alternating current) equivalent.

Resistance is due to electrons in a conductor colliding with the ionic lattice of the conductor meaning that electrical energy is converted into heat. Different materials have different resistivities (a property defining how resistive material of given dimensions will be).

However, when considering AC you must remember that it oscillates like a sine wave so the sign is always changing. This means that other effects need to be considered – namely inductance and capacitance.

Inductance is most obvious in the coiled wire. When a current flows through a wire a circular magnetic field is created around it. If you coil the wire into a solenoid the fields around the wire sum up and you get a magnetic field similar to that of a bar magnet on the outside but you get a uniform magnetic field on the inside. With AC since the sign is always changing the direction of the field in the wires is always changing – so the magnetic field of the solenoid is also changing all the time. Now when field lines cut across a conductor an emf is generated in such a way to reduce the effects that created it (this is a combination of Lenz's and Faraday's laws which state mathematically that E = N*d(thi)/dt , where thi is the magnetic flux linkage). This means that when an AC flows through a conductor a small back emf or back current is induced reducing the overall current.

Capacitance is a property best illustrated by two metal plates separated by an insulator (which we call a capacitor). When current flows electrons build up on the negative plate. An electric field propagates and repels electrons on the opposite plate making it positively charged. Due to the build-up of electrons on the negative plate, incoming electrons are also repelled so the total current eventually falls to zero in exponential decay. The capacitance is defined as the charge stored/displaced across a capacitor divided by the potential difference across it and can also be calculated by the size of the plates and the primitivity of the insulator.

So, simply resistance and impedance have different fundamental origins even though the calculation for their value is the same:

$$R = V/I$$

Impedance is a more general term for resistance that also includes reactance.

In other words, resistance is the opposition to a steady electric current. Pure resistance does not change with frequency, and typically the only time only resistance is considered is with DC (direct current – not changing) electricity.

Reactance, however, is a measure of the type of opposition to AC electricity due to capacitance or inductance. This opposition varies with frequency. For example, a capacitor only allows DC to flow for a short while until it is charged; at that point, the current will stop flowing and it will look like an open. However, if a very high frequency is put across that capacitor (a signal that has a voltage which is changing very quickly back and forth), the capacitor will look like a short circuit. The capacitor has a reactance which is inversely proportional to frequency. An inductor has a reactance which is directly proportional to frequency DC flows through easily while high-frequency AC is stopped.

Impedance is the total contribution of both resistance and reactance.

- **Ques: 88. What is the difference between resistance and reactance?**

Solution:

Resistance is the measure of opposition to the current flow offered by the material. Usually denoted by R.

Reactance is the resistance offered to the AC currents by inductors and capacitors only. Usually denoted by X.

For capacitors $X = 1/(2\pi fC)$, where f is the frequency and C is the capacitance.

For inductors $X = 2\pi fL$, where f is the frequency and L is the inductance.

Impedance is the sum of the resistance and reactance of a circuit

denoted by $Z = R + jX$ (for primarily inductive circuits) or $Z = R - jX$ (for PRIMARILY capacitive circuits).

where $j = \sqrt{(-1)}$.

There are two major differences:

1. Resistance is independent of the frequency of input signal, whereas reactance is a quantity that depends on the frequency of the input

signal. (Inductive reactance is directly proportional to the frequency and capacitive reactance is inversely proportional to the frequency.)

2. Reactance is a wattless quantity. Energy is stored in the form of energy and can be used again and therefore the power loss is 0. Resistance is wattfull quantity. A resistor converts the energy it receives and dissipates it in the form of heat.

- **Ques: 89. What is the difference between reactance and reluctance?**

Solution:
Reluctance is a unit measuring the opposition to the flow of magnetic flux within magnetic materials and is analogous to resistance in electrical circuits. Capacitance, C, is measured in Farads and has reactance given by $X = 1/(2\pi fC)$.

The reactance is dependent upon frequency.

- **Ques: 90. What is the difference between resistance and inductance?**

Solution:
R: resistance is that property of a material which opposes the flow of current.
L: inductance is that property of a material which resists the change in current.

- **Ques: 91. What are the values of resistance, reactance, and impedance of resistor, capacitor, and inductor in series RLC circuit?**

Solution:
In series-RLC circuit: A resistance, a capacitance and an inductance connected in series across an alternating supply are connected as:

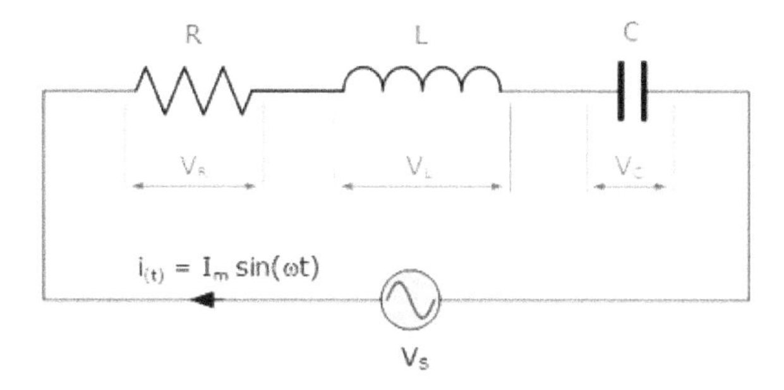

In a pure ohmic resistor, the voltage waveforms are "in-phase" with the current. In a pure inductance the voltage waveform "leads" the current by 90°, giving us the expression of ELI. In a pure capacitance the voltage waveform "lags" the current by 90°, giving us the expression of ICE.

This phase difference, Φ depends upon the reactive value of the components being used and hopefully by now we know that reactance, (X) is zero if the circuit element is resistive, positive if the circuit element is inductive and negative if it is capacitive thus giving their resulting impedances as:

Circuit Element	Resistance (R)	Reactance (X)	Impedance (Z)
Resistor	R	O	$Z_R = R$ $= R\angle 0°$
Inductor	O	ωL	$Z_L = j\omega L$ $= \omega L\angle + 90°$
Capacitor	O	$\frac{1}{\omega C}$	$Z_C = \frac{1}{j\omega C}$ $= \frac{1}{\omega C}\angle - 90°$

The phasor diagram is represented as

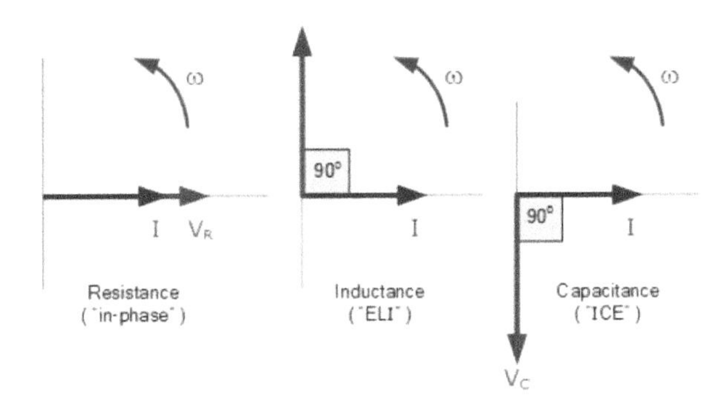

- **Ques: 92. Why current leads in the capacitor?**

Solution:
We know for a capacitor that

$$Q = CV$$

where Q is the charge on the capacitor's plates, C is its capacitance, and V is the voltage across the capacitor.

We also know that I, the electric current is the flow of electric charge with time:

$$I = dQ/dt$$

Combine these two, and for a capacitor, we see:

$$I = dQ/dt = C*dV/dt$$

Now, if we have a sinusoidal input voltage, we can calculate the current across the capacitor as a function of the voltage:

$$V(t) = \sin(t)$$
$$I(t) = C*dV(t)/dt = C*\cos(t)$$

But cos(t) is just sin(t) plus pi/2 radians (90 degrees). So our final equations for the capacitor circuit above become:

$$V(t) = \sin(t)$$
$$I(t) = C*\cos(t) = C*\sin(t + \pi/2)$$

So, for a sinusoidal input voltage, we see that we also get a sinusoidal current, but the current leads the voltage by pi/2 radians (90 degrees).

- **Ques: 93. Why current lags by voltage by 90 degrees in the inductor?**

Solution:
Inductor: Inductors are storage elements that store energy in the form of magnetic fields. The current flowing through the inductor is the reason for the establishment of the magnetic field.

Lenz's law states that the effect tends to oppose the cause. Try to follow my reasoning here. A voltage was applied across the inductor. This causes current to flow through it. The current passing through the inductor establishes a magnetic field (corkscrew rule). According to Faraday's law, when a current-carrying conductor interacts with a magnetic field, there is an EMF induced in the conductor. So, the current flowing through the inductor and the magnetic field established due to the current interact with each other to induce an EMF in the inductor, which opposes the applied voltage (effect opposes cause) according to Lenz's law. This EMF causes a current to flow, which opposes the main current. This causes a reduction in the main current.

Another way of looking at it is that if you apply your source voltage to an inductor,
$V(S) = \sin(wt)$ then the current must be

$$I = \sin(wt)/R$$

So, by the inductor formula:

$$V(L) = L\ di/dt = L/R\ d\ \sin(wt)/dt = L/R\ \cos(wt)$$

And we know that
cos(wt) is 90 degrees ahead of sin(wt).

- **Ques: 94. What is the difference between EMF and voltage?**

Solution:
The EMF is measured between the endpoint of the source, when no current flow through it, whereas, the voltage is measured between any two points of the closed circuit. The EMF is generated by the electrochemical cell, dynamo, photodiodes, etc., whereas the voltage is caused by the electric and magnetic fields.

- **Ques: 95. What do you mean by the bathtub curve?**

Solution:
The bathtub curve is widely used in reliability engineering. It describes a particular form of the hazard function which comprises three parts:

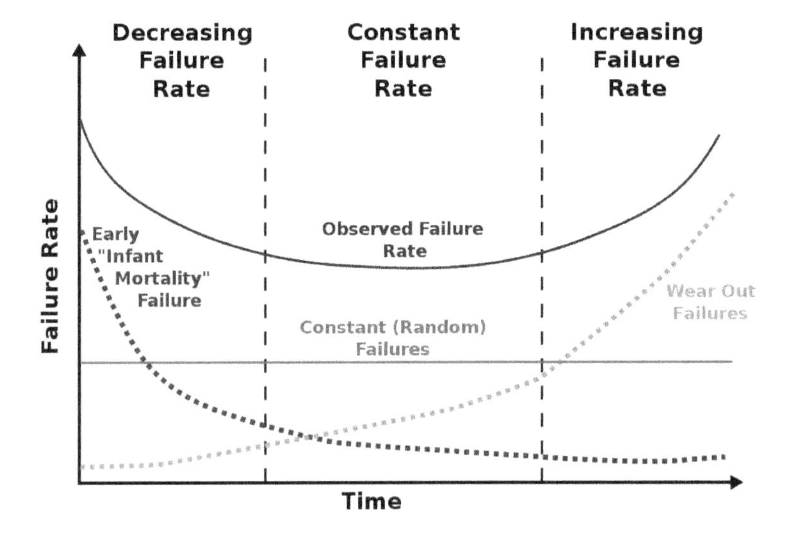

The first part is a decreasing failure rate, known as early failures.
The second part is a constant failure rate, known as random failures.
The third part is an increasing failure rate, known as wear-out failures.
The name is derived from the cross-sectional shape of a bathtub: steep sides and a flat bottom.

- **Ques: 96. What do you mean by MTBF, MTTR, and MTTF?**

Solution:

Mean time between failure (MTBF) is a reliability term used to provide the number of failures per million hours for a product. This is the most common inquiry about a product's life span and is important in the decision-making process of the end user. MTBF is more important for industries and integrators than for consumers. Most consumers are price driven and will not consider MTBF, nor is the data often readily available. On the other hand, when equipment such as media converters or switches must be installed into mission-critical applications, MTBF becomes very important. Also, MTBF maybe an expected line item in an RFQ (request for quote). Without the proper data, a manufacturer's piece of equipment would be immediately disqualified.

Mean time to repair (MTTR) is the time needed to repair a failed hardware module. In an operational system, repair generally means replacing a failed hardware part. Thus, hardware MTTR could be viewed as a mean time to replace a failed hardware module. Taking too long to repair a product drives up the cost of the installation in the long run, due to downtime until the new part arrives and the possible window of time required to schedule the installation. To avoid MTTR, many companies purchase spare products so that a replacement can be installed quickly. Generally, however, customers will inquire about the turnaround time of repairing a product, and indirectly, that can fall into the MTTR category.

Mean time to failure (MTTF) is a basic measure of reliability for nonrepairable systems. It is the mean time expected until the first failure of a piece of equipment. MTTF is a statistical value and is meant to be the mean over a long period and a large number of units. Technically, MTBF should be used only about a repairable item, while MTTF should be used for nonrepairable items. However, MTBF is commonly used for both repairable and nonrepairable items.

$$MTBF = MTTR + MTTF$$

- **Ques: 97. What is the significance of FIT?**

Solution:

Failure in time (FIT) is a way of reporting MTBF. FIT reports the number of expected failures per one billion hours of operation for a device. This term is used particularly by the semiconductor industry but is also used by component manufacturers. FIT can be quantified in many ways: 1000 devices for 1 million hours or 1 million devices for 1000 hours each, and other combinations. FIT and CL (confidence limits) are often provided together. In common usage, a claim to 95% confidence in something is normally taken as indicating virtual certainty. In statistics, a claim to 95% confidence simply means that the researcher has seen something occur that only happens one time in 20 or less. For example, component manufacturers will take a small sampling of a component, test x number of hours, and then determine if there were any failures in the testbed. Based on the number of failures that occur, the CL will then be provided as well.

- **Ques: 98. Name any five VLSI companies.**

Solution:

1. ST Microelectronics
2. Synopsys
3. XILINX
4. Cadence Design System
5. Intel

- **Ques: 99. What are the characteristics of an ideal opamp?**

Solution:

1. Infinite open-loop gain
2. Infinite input impedance
3. Zero output impedance
4. Infinite frequency bandwidth
5. The infinite common-mode rejection ratio

- **Ques: 100. What do you mean by CMRR and slew rate?**

Solution:

CMMR stands for common mode rejection ratio and it is defined as the ratio of differential voltage gain to the common-mode voltage gain

$$\text{CMMR} = \text{Ad/Ac}$$

where Ad is differential voltage gain and Ac is a common-mode voltage gain

The slew rate of an op-amp or an amplifier circuit is the rate of change in the output voltage caused by a step change on the input. It is measured as a voltage change in a given time – typically V/μs or V/ms.

- **Ques: 101. What do you mean by the transient response and steady-state response?**

Solution:

In a system, when certain input changes, it takes a while for the output to stabilize and reach its final state. This interim phase is called the transient phase. The final state is the steady state and the system will stay there indefinitely until some input changes again.

- **Ques: 102. Draw voltage transfer curve of OpAmp.**

Solution:
This means that the output voltage is directly proportional to the input difference voltage only until it reaches the saturation voltages and thereafter the output voltage remains constant. Thus curve is called an ideal voltage transfer curve, ideal because output offset voltage is assumed to be zero.

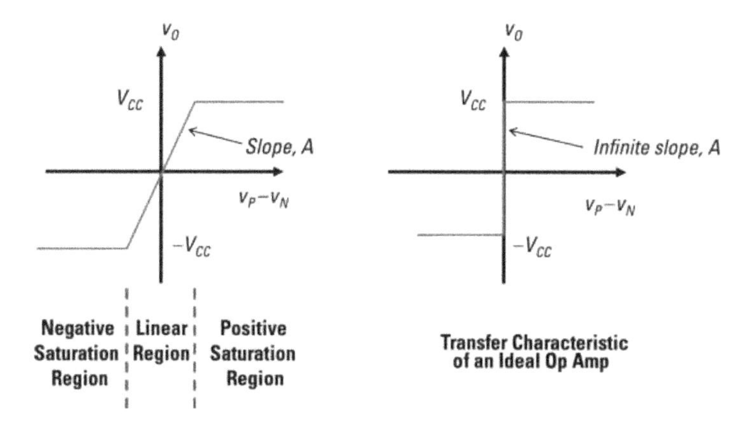

- **Ques: 103. What do you mean by input offset voltage?**

Solution:
It is the voltage that is applied between the input and output terminal of OpAmp to balance the amplifier and null the output voltage. The input offset voltage is defined as the voltage that must be applied between the two input terminals of the op-amp to obtain zero volts at the output.

- **Ques: 104. Differentiate between microprocessor and microcontroller.**

Solution:
1. The major difference in both of them is the presence of external peripheral, where microcontrollers have RAM, ROM, and EEPROM embedded in it while we have to use external circuits in case of microprocessors.

2. As all the peripheral of the microcontroller is on the single chip it is compact while the microprocessor is bulky.

3. Microcontrollers are made by using complementary metal–oxide–semiconductor technology, so they are far cheaper than microprocessors. Also, the applications made with microcontrollers are cheaper because they need lesser external components, while the overall cost of systems made with microprocessors is high because of the high number of external components required for such systems.

4. The processing speed of microcontrollers is about 8–50 MHz, but in contrary processing speed of general microprocessors is >1 GHz so it works much faster than microcontrollers.

5. Generally, microcontrollers have a power-saving system, like idle mode or power-saving mode so overall it uses less power and also since external components are low overall consumption of power is less. While in microprocessors generally there is no power saving system and also many external components are used with it, so its power consumption is high in comparison with microcontrollers.

6. Microcontrollers are compact so it makes them a favorable and efficient system for small products and applications while microprocessors are bulky so they are preferred for larger applications.

7. Tasks performed by microcontrollers are limited and generally less complex. While tasks performed by microprocessors are software development, game development, website, document-making, etc., which are generally more complex so require more memory and speed so that's why external ROM and RAM are used with it.

8. Microcontrollers are based on Harvard architecture where program memory and data memory are separate while microprocessors are based on the von Neumann model where program and data are stored in the same memory module.

Arduino UNO is a microprocessor board based on ATmega328P.

- **Ques: 105. What do you know about MOSFET?**

Solution:
The MOSFET is a metal–oxide–semiconductor field-effect transistor, which is controlled by voltage rather than current. It works electronically, by varying the width of the channel along with the charge carriers flow. The wider the channel, the better the device conductance.

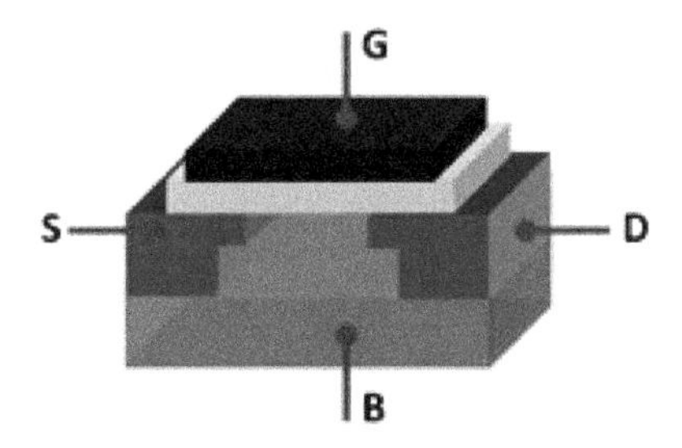

The charge carriers enter through the source and exit through the drain. The width of the channel is controlled by the voltage on channel, i.e., gate, which is located physically in between the source and drain and insulated from the channel by an extremely thin layer of metal oxide.

- **Ques: 106. Explain how MOSFET functions.**

Solution:

1. Depletion mode

When there is no voltage on the gate, the channel shows its maximum conductance. As the voltage on the gate is either positive or negative, the channel conductivity decreases.

2. Enhancement mode

When there is no voltage on the gate the device does not conduct. More is the voltage on the gate, the better the device can conduct.

- **Ques: 107. Draw the characteristics of depletion and enhancement type NMOS and PMOS.**

Solution:

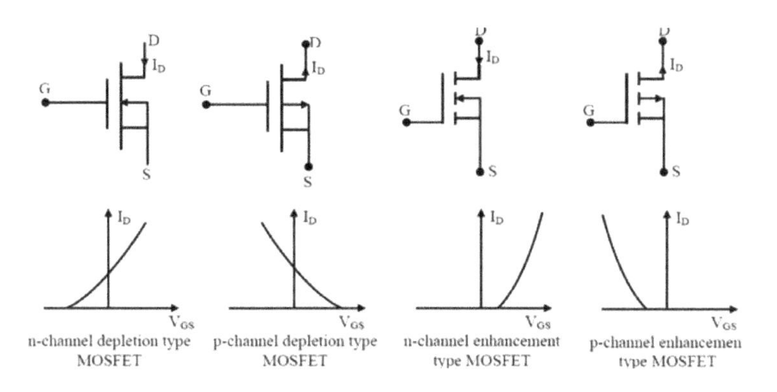

• **Ques: 108. Draw I–V characteristics of MOSFET.**

Solution:

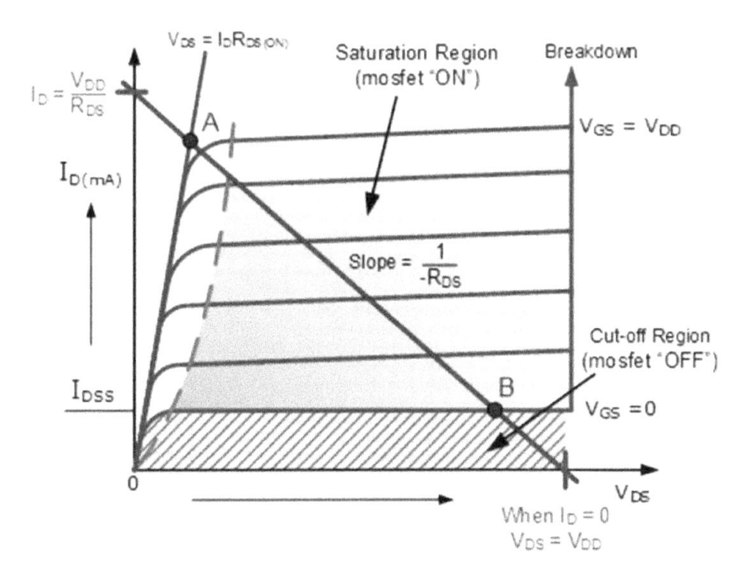

• **Ques: 109. What do you understand by pinch-off voltage?**

Solution:

Pinch-off voltage: Pinch-off voltage is the drain to source voltage after which the drain to source current becomes almost constant and JFET enters into saturation region and is defined only when the gate to source voltage is zero.

• **Ques: 110. Explain MOSFET as a switch.**

Solution:

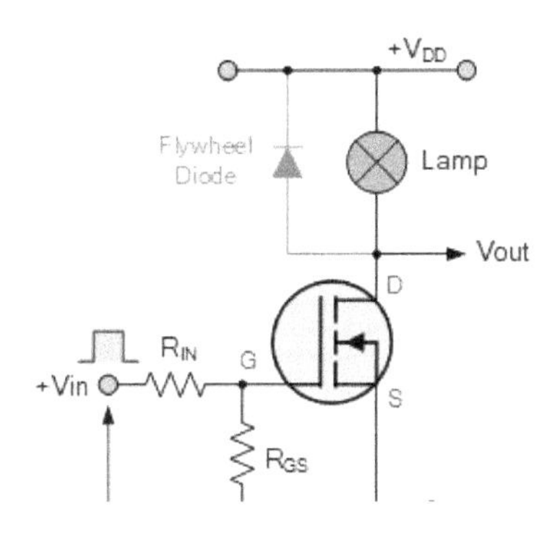

In this circuit arrangement, an enhanced mode and N-channel MOSFET are being used to switch a sample lamp ON and OFF. The positive gate voltage is applied to the base of the transistor and the lamp is ON (VGS=+v) or at zero voltage level the device turns off (VGS=0). If the resistive load of the lamp was to be replaced by an inductive load and connected to the relay or diode which is protected to the load. In the above circuit, it is a very simple circuit for switching a resistive load such as a lamp or an LED. But when using MOSFET to switch either inductive load or capacitive load protection is required to contain the MOSFET device. If we do not give protection the MOSFET device would get damaged. For the MOSFET to operate as an analog switching device, it needs to be switched between its cut-off region where VGS = 0 and saturation region where VGS = +v.

- **Ques: 111. What is forward transconductance in MOSFET?**

Solution:
The forward transconductance is the ratio of Id and (Vgs–Vgs(th)). In the MOSFET switching circuit, it determines the clamping voltage level of gate-source voltage and thus influences the $\frac{d}{dt}V.DS$ during turn-on and turn-off.

- **Ques: 112. What are the three regions of operation in MOSFET?**

Solution:
The three operational regions of MOSFET are the following:

1. Cut-off Region
With VGS $<$ Vthreshold the gate-source voltage is much lower than the transistors threshold voltage so the MOSFET transistor is switched "fully-OFF" thus, ID = 0, with the transistor acting as an open switch regardless of the value of VDS.

2. Linear (Ohmic) Region
With VGS $>$ Vthreshold and VDS $<$ VGS, the transistor is in its constant resistance region behaving as a voltage-controlled resistance whose resistive value is determined by the gate voltage, VGS level. It is also known as the triode region.

3. Saturation Region
With VGS $>$ Vthreshold and VDS $>$ VGS, the transistor is in its constant current region and is, therefore "fully-ON." The drain current ID = maximum with the transistor acting as a closed switch.

- **Ques: 113. Draw I–V characteristics of BJT.**

Solution:

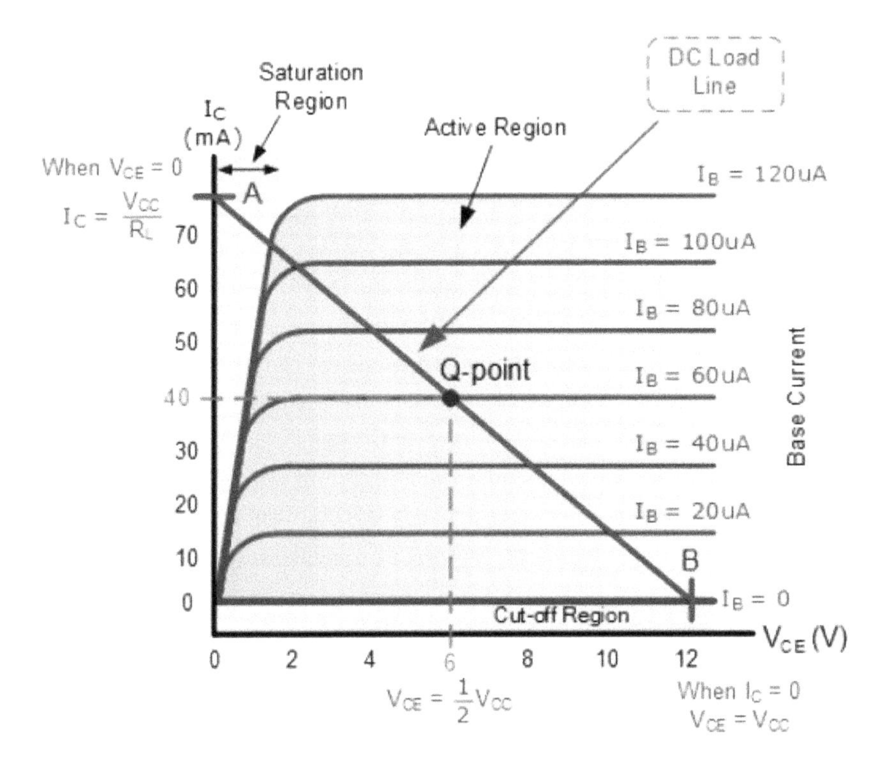

Q-point is an acronym for the quiescent point. Q-point is the operating point of the transistor (ICQ, VCEQ) at which it is biased. The concept of Q-point is used when the transistor act as an amplifying device and hence is operated in the active region of input–output characteristics.

- **Ques: 114. What do you mean by Zener diode?**

Solution:

A Zener diode is a type of diode that allows current to flow not only from its anode to its cathode but also in the reverse direction when the Zener voltage is reached. Zener diodes have a highly doped p–n junction.

The below diagram shows the V–I characteristics of the Zener diode behavior. When the diode is connected in forward bias diode acts as a normal diode. When the reverse bias voltage is greater than a predetermined voltage then the Zener breakdown voltage occurs.

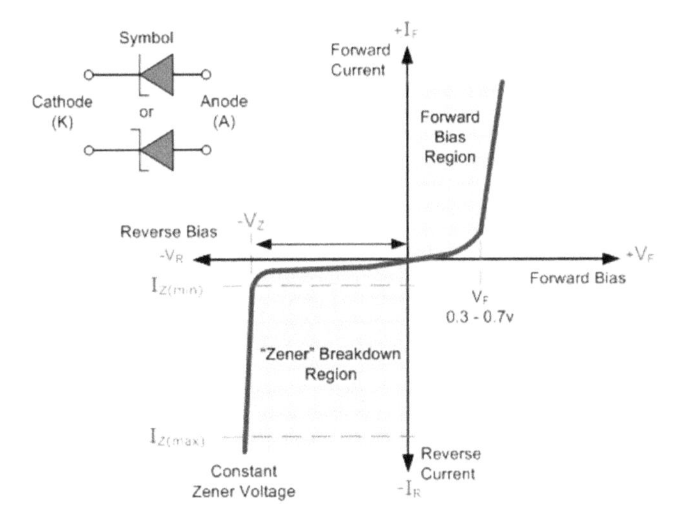

- **Ques: 115. What do you mean by Schmitt trigger?**

Solution:
It is a bistable circuit in which the output increases to a steady maximum when the input rises above a certain threshold and decreases almost to zero when the input voltage falls below another threshold.

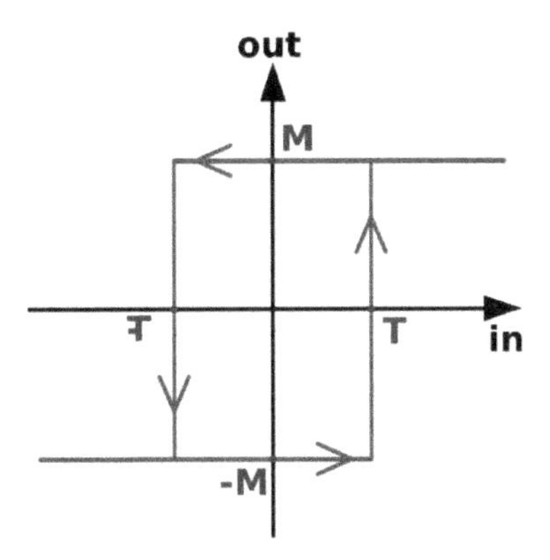

Schmitt trigger is an electronic circuit with positive feedback which holds the output level until the input signal to the comparator is higher than the threshold. It converts a sinusoidal or any analog signal to a digital signal.

- **Ques: 116. Draw the CE, CB, and CC configuration of BJT?**

Solution:

Basic circuit	Common emitter	Common collector	Common base
Voltage gain	high	less than unity	high, same as CE
Current gain	high	high	less than unity
Power gain	high	moderate	moderate
Phase inversion	yes	no	no
Input impedance	moderate ≃ 1 k	highest ≃ 300 k	low ≃ 50 Ω
Output impedance	moderate ≃ 50 k	low ≃ 300 Ω	highest ≃ 1 Meg

The output characteristics of the CE configuration are

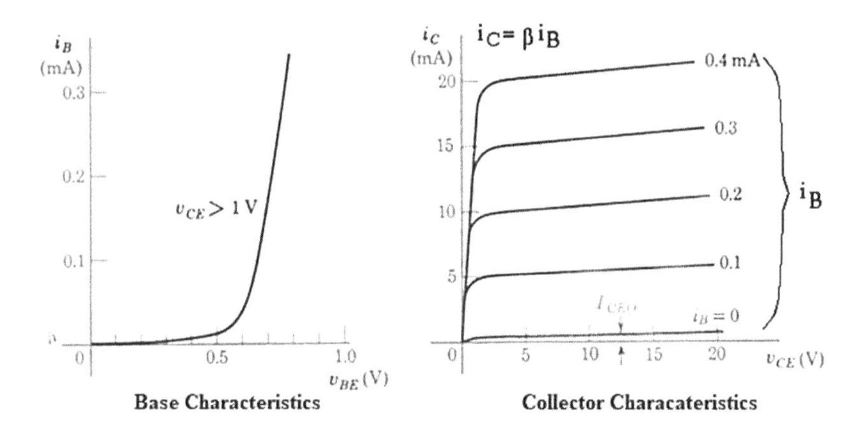

- **Ques: 117. What is the relation between α and β in the case of BJT?**

Solution:

The current gain or current transfer ratio of this CB circuit, denoted by α is defined as the ratio between collector current treated as the output and the emitter current treated as the input:

$$\alpha = \frac{I_C}{I_E} < 1, \quad \text{e.g. } \alpha = 99\% \approx 1$$

The current gain of the CE circuit, denoted by β, is defined as the ratio between the collector current treated as the output and the base current treated as the input:

$$\beta = \frac{I_C}{I_B} = \frac{\alpha I_E}{(1-\alpha)I_E} = \frac{\alpha}{1-\alpha}$$

- **Ques: 118. Explain validation and verification in VLSI design.**

Solution:
In simple terms verification is presilicon and validation is postsilicon.

Verification is generally used for frontend, i.e., the actual verification of the RTL, which can be done mainly using System Verilog/Verilog and using a methodology like OVM, UVM (where we create the entire test environment, which can include transactor, scoreboard, monitor, packets).

The most important thing to note here is to understand that in the verification step, the input to a verification engineer is specification and he/she checks if the RTL designer has coded the given specification accordingly.

Validation terminology is used once the silicon is back from the lab and the team intends to check if the chip is fabricated well and is still functions as it was supposed to before it went for fabrication with no error.

The most important thing to note here is that here the input is not from specifications, i.e., we are not validating the specifications and the chip, but instead, we assume that the verification engineer has done his/her part well, and hence we check here if the device is fabricated as per what was given to the fab thus validating the silicon.

- **Ques: 119. What do you understand by black-box testing?**

Solution:
Testing based on an analysis of the specification of a piece of software without reference to its internal workings. The goal is to test how well the component conforms to the published requirements for the component.

- **Ques: 120. What is the difference between a pilot and beta testing?**

Solution:
The difference between a pilot and beta testing is that pilot testing is nothing but using the product (limited to some users) and in beta testing, we do not

input real data, but it's installed at the end customer to validate if the product can be used in production.

- **Ques: 121. What's the difference between system testing and acceptance testing?**

Solution:
Acceptance testing checks the system against the "requirements." It is similar to system testing in that the whole system is checked but the important difference is the change in focus. The system testing checks that the system that was specified has been delivered. Acceptance testing checks that the system will deliver what was requested. The customer should always do acceptance testing and not the developer.

- **Ques: 122. Explain the term "defect?"**

Solution:
The variation between the actual results and expected results is known as a defect.

If a developer finds an issue and corrects it by himself in the development phase then it's called a defect.

- **Ques: 123. Explain the term "bug?"**

Solution:
If testers find any mismatch in the application/system in the testing phase then they call it bug.

- **Ques: 124. Explain the term "error?"**

Solution:
We can't compile or run a program due to coding mistakes in a program. If a developer unable to successfully compile or run a program then they call it an error.

- **Ques: 125. Explain the term "failure?"**

Solution:
Once the product is deployed and customers find any issues then they call the product as a failure product. After release, if an end user finds an issue then that particular issue is called a failure.

- **Ques: 126. What are the different types of "defect?"**

Solution:
Severity defines the degree of impact. Thus, defect's severity reflects the degree or intensity of a particular defect, to impact a software product or

its working, adversely. Based on the severity metric, a defect maybe further categorized into the following:

- Critical: The defects termed as "critical," needs immediate attention and treatment. A critical defect directly affects the critical and essential functionalities, which may affect a software product or its functionality on a large scale, such as failure of a feature/functionality or the whole system, system crash-down, etc.
- Major: Defects, which are responsible for affecting the core and major functionalities of a software product. Although, these defects do not result in complete failure of a system, but may bring several major functions of the software to rest.
- Minor: These defects produce minor impact, and does not have any significant influence on a software product. The results of these defects maybe seen in the product's working; however, it does not stop users to execute a task, which maybe carried out using some other alternative.
- Trivial: These types of defects, have no impact on the working of a product, and sometimes, it is ignored and skipped, such as spelling or grammatical mistakes.

- **Ques: 127. What is the difference between quality assurance, quality control, and testing?**

Solution:

Quality assurance is the process of planning and defining the way of monitoring and implementing the quality (test) processes within a team and organization. This method defines and sets the quality standards of the projects.

Quality control is the process of finding defects and providing suggestions to improve the quality of the software. The methods used by quality control are usually established by quality assurance.

It is the primary responsibility of the testing team to implement quality control.

Testing is the process of finding defects/bugs. It validates whether the software built by the development team meets the requirements set by the user and the standards set by the organization.

Here the main focus is on finding bugs and testing teams acts as a quality gatekeeper.

- **Ques: 128. Why power stripes routed in the top metal layers?**

Solution:

The resistivity of top metal layers is less and hence less IR drop is seen in the power distribution network. If power stripes are routed in lower metal layers this will use a good amount of lower routing resources and therefore it can create routing congestion.

- **Ques: 129. What are several factors to improve the propagation delay of a standard cell?**

Solution:
Improve the input transition to the cell under consideration by upsizing the driver.

Reduce the load seen by the cell under consideration, either by placement refinement or buffering.

If allowed increase the drive strength or replace it with LVT (low threshold voltage) cell.

- **Ques: 130. What are the various ways of timing optimization in synthesis tools?**

Solution:

1. Logic optimization: buffer sizing, cell sizing, level adjustment, dummy buffering, etc.
2. Less number of logics between flip-flops speedup the design.
3. Optimize drive strength of the cell, so it is capable of driving more load and hence reducing the cell delay.
4. Better selection of designware components (select timing optimized designware components).
5. Use LVT and SVT (standard threshold voltage) cells if allowed.

- **Ques: 131. What are the various techniques to resolve congestion/noise?**

Solution:
Routing and placement congestion all depend upon the connectivity in the netlist, a better floor plan can reduce the congestion.

Noise can be reduced by optimizing the overlap of nets in the design.

- **Ques: 132. What is PVT analysis of circuits and how it can be performed?**

Solution:
Process voltage temperature variation analysis is important for analyzing model random mismatch of various parameters like length and width of transistors, supply voltage, the thickness of the oxide, temperature, and threshold voltage. It can be done using Monte Carlo analysis using Cadence.

- **Ques: 133. Write the steps for Monte Carlo analysis.**

Solution:

1. Open the circuit schematic in Cadence Virtuoso.
2. Go to “File” and click on “ADEXL.”
3. Click on “Open Existing” even if you haven’t done this analysis before. A new window opens by default, but if you have an existing view, you will not overwrite it.

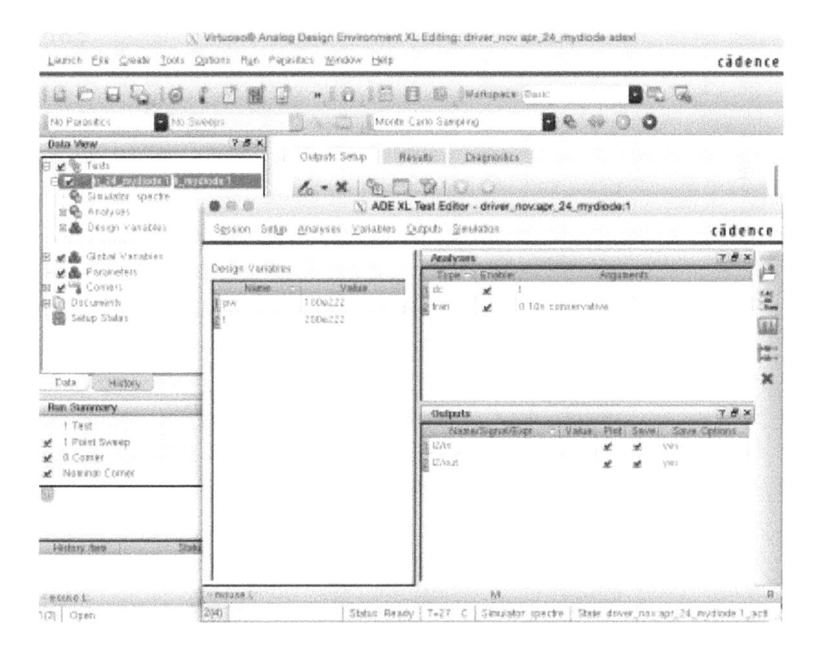

4. Click on “Tests” and click on “Click to add test.” There are a couple of other ways to add a new test:
   Click on “Create” and then “Test.”
   Click on the yellow icon in the toolbar.
5. You can pick the whole schematic to run the analysis or one particular component.
6. To add the test, if you have a previously saved analysis click on “Session” and “Load state.” To add a new test, click on “Setup,” “Model libraries” and add the library files required. Usually, libraries are written in scheme language. If you have an academic library, you will have the files in the spectre folder. You will have to include “/models/all designs.scs’ and ‘/models/design.scs.”

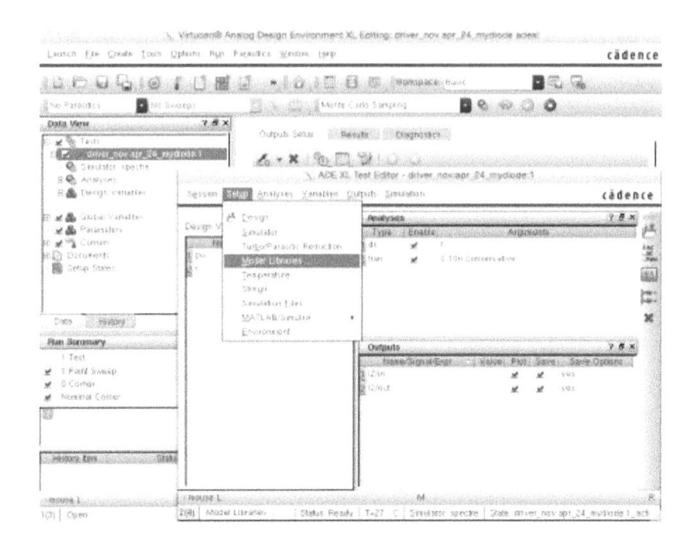

7. To add design variables, right click in the design variable window and click on "Copy from cell view."
8. Click on analysis, then choose and select the analysis required. I chose DC and transient analysis.
9. Click on "Outputs," "To be plotted," and then "Select on schematic." Select the outputs to be plotted.
10. Save the session and go to the ADEXL window again.
11. Click on the green icon beside the drop-down menu saying "Monte Carlo Sampling." A window will open as shown in the picture.

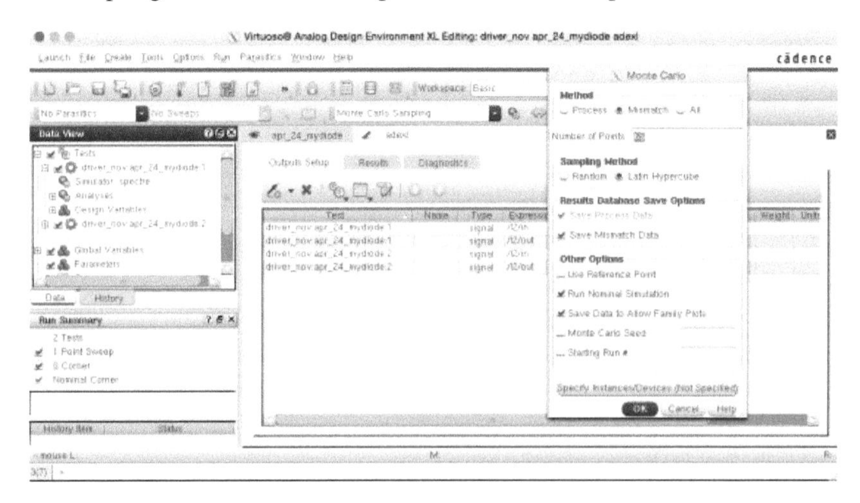

12. Enter the number of points, For example, ran the analysis 65 times.
13. Select "Run nominal simulation" to run a test analysis in addition to the number of runs selected.
14. Click "OK" to start the simulation.
15. Once the simulation is complete you will see the graph window appear with all the results.

- **Ques: 134. What are the various types of jobs for VLSI engineers?**

Solution:

1. **Design Engineer** Takes specifications, defines architecture, does circuit design, runs simulations, supervises layout, tapes out the chip to the foundry, evaluate the prototype once the chip comes back from the fab.
2. **Product Engineer**
   Gets involved in the project during the design phase, ensures manufacturability, develops characterization plan, assembly guidelines, develops quality and reliability plan, evaluates the chip with the design engineer, evaluates the chip through characterization, reliability qualification and manufacturing yield point of view (statistical data analysis). He/she is responsible for production release and is therefore regarded as a team leader on the project. Postproduction, he/she is responsible for customer returns, failure analysis, and corrective actions including design changes.
3. **Test Engineer** Develops test plan for the chip based on specifications and datasheet, creates characterization and production program for the bench test or the ATE (automatic test equipment), designs test board hardware, correlates ATE results with the bench results to validate silicon to compare with simulation results.
4. **Applications Engineer** Defines new products from a system point of view at the customers' end, based on marketing input. His/her mission is to ensure the chip works in the system designed or used by the customers and complies with appropriate standards (such as Ethernet, SONET, WiFi, etc.). He/she is responsible for all customer technical support, firmware development, evaluation boards, datasheets, etc.
5. **Process Engineer** This is a highly specialized function that involves new wafer process development, device modeling, and lots of research and development projects.
6. **Packaging Engineer** This is another highly specialized job function. He/she develops precision packaging technology, new package designs for the chips, does the characterization of new packages, and modeling.

7. **CAD Engineer** This is an engineering function that supports the design engineering function. He/she is responsible for acquiring, maintaining, or developing all CAD tools used by a design engineer. Most companies buy commercially available CAD tools for schematic capture, simulation, synthesis, test vector generation, layout, parametric extraction, power estimation, and timing closure; but in several cases, these tools need some type of customization. A CAD engineer needs to be highly skilled in the use of these tools, be able to write software routines to automate as many functions as possible and have a clear understanding of the entire design flow.

- **Ques: 135. What do you mean by an oscillator?**

Solution:
An electronic oscillator is an electronic circuit that produces a periodic, oscillating electronic signal, often a sine wave or a square wave. Oscillators convert DC from a power supply to an AC signal. They are widely used in many electronic devices.

- **Ques: 136. What is a ring oscillator?**

Solution:
A ring oscillator is a device composed of an odd number of NOT gates in a ring, whose output oscillates between two voltage levels, representing true and false. The NOT gates, or inverters, are attached in a chain and the output of the last inverter is fed back into the first.

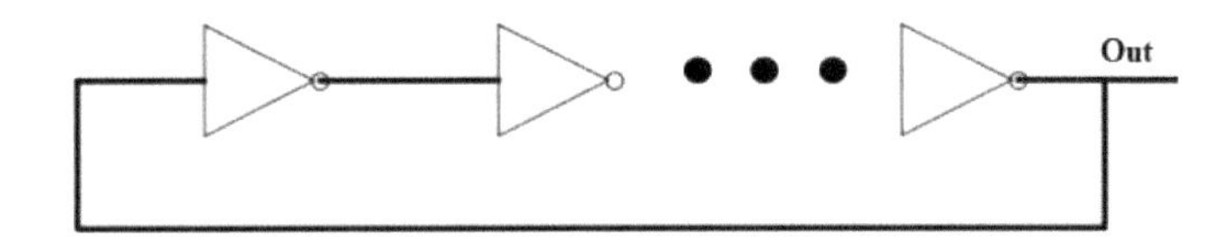

The total period of operation is the product of 2*number of gates and gate (inverter) delay. And the frequency of operation will be inverse of period.

It is used as prototype circuits for modeling and designing new semiconductor processes due to simplicity in design and ease of use. It also forms a part of the clock recovery circuit.

- **Ques: 137. What is the basic difference between analog and digital design?**

Solution:
Digital design is distinct from analog design. In analog circuits, we deal with physical signals which are continuous in amplitude and time. Ex: biological data, sensor output, audio, video, etc.

Analog design is quite challenging than digital design as analog circuits are sensitive to noise, operating voltages, loading conditions, and other conditions that have severe effects on performance. Even process technology poses certain topological limitations on the circuit. Analog designer has to deal with real-time continuous signals and even manipulate them effectively even in a harsh environment and in brutal operating conditions.

Digital design, on the other hand, is easier to process and has great immunity to noise. No room for automation in analog design as every application requires a different design. However, digital design can be automated. Analog circuits generally deal with the instantaneous value of voltage and current (real-time). It can take any value within the domain of specifications for the device and consists of passive elements that contribute to the noise (thermal) of the circuit. They are usually more sensitive to external noise more so because for a particular function in an analog design uses lot fewer transistors providing design challenges over process corners and temperature ranges. An analog design deals with a lot of device-level physics and the state of the transistor plays a very important role. Digital circuits on the other hand deal with only two logic levels 0 and 1 (Is it true that according to quantum mechanics there is a third logic level?), deal with lot more transistors for a particular logic, easier to design complex designs, flexible logic synthesis, and greater speed although at the cost of greater power. Less sensitive to noise. Design and analysis of such circuits is dependent on the clock. However, the challenge lies in negating the timing and load delays and ensuring there is no setup or hold violation.

- **Ques: 138. What is stuck at fault?**

Solution:
A stuck at fault is a particular fault model used by fault simulators and automatic test pattern generation (ATPG) tools to mimic a manufacturing defect within an IC. Individual signals and pins are assumed to be stuck at logical "1," "0," and "X." For example, an output is tied to a logical 1 state during test generation to assure that a manufacturing defect with that type of behavior can be found with a specific test pattern. Likewise, the output could be tied to a logical 0 to model the behavior of a defective circuit that cannot switch its output pin.

- **Ques: 139. What is physical verification?**

Solution:
Physical verification of the design, involves DRC (design rule check), LVS (layout versus schematic) Check, XOR Checks, ERC (electrical rule check), and antenna checks.

☐ XOR Check

This step involves comparing two layout databases/GDS by XOR operation of the layout geometries. This check results in a database that has all the mismatching geometries in both the layouts. This check is typically run after a metal spin, wherein the respin database/GDS is compared with the previously taped out database/GDS.

☐ Antenna Check

Antenna checks are used to limit the damage of the thin gate oxide during the manufacturing process due to charge accumulation on the interconnect layers (metal, polysilicon) during certain fabrication steps like plasma etching, which creates highly ionized matter to etch. The antenna is a metal interconnect, i.e., a conductor such as polysilicon or metal, that is not electrically connected to silicon or grounded, during the processing steps of the wafer. If the connection to silicon does not exist, charges may build up on the interconnect to the point at which rapid discharge does take place and permanent physical damage results in the thin transistor gate oxide. This rapid and destructive phenomenon is known as the antenna effect. The antenna ratio is defined as the ratio between the physical area of the conductors making up the antenna to the total gate oxide area to which the antenna is electrically connected.

☐ ERC

ERC involves checking a design for all well and substrate areas for proper contacts and spacings thereby ensuring correct power and ground connections. ERC steps can also involve checks for unconnected inputs or shorted outputs.

- **Ques: 140. What are DRC and LVS?**

Solution:
DRC and LVS are verification processes. Reliable device fabrication at modern deep submicrometre (0.13 μm and below) requires strict observance of transistor spacing, metal layer thickness, and power density rules. DRC exhaustively compares the physical netlist against a set of "foundry design rules" (from the foundry operator), then flags any observed violations.

DRC

It determines whether the layout of a chip satisfies a series of recommended parameters called design rules. Design rules are a set of parameters provided by semiconductor manufacturers to the designers, to verify the correctness of a mask set. It varies based on the semiconductor manufacturing process. This ruleset describes certain restrictions in geometry and connectivity to ensure that the design has sufficient margin to take care of any variability in the manufacturing process.

Design rule checks are nothing but physical checks of metal width, pitch, and spacing requirement for the different layers concerning different manufacturing processes. If we give physical connection to the components without considering the DRC rules, then it will lead to failure of the functionality of the chip, so all DRC violations have to be cleaned up.

After the completion of physical connection, we check every polygon in the design, based on the design rules and report all the violations. This whole process is called the DRC.

Typical DRC rules are the following:

☐ Interior
☐ Exterior
☐ Enclosure
☐ Extension

LVS

LVS is a process that confirms that the layout has the same structure as the associated schematic; this is typically the final step in the layout process. The LVS tool takes as an input a schematic diagram and the extracted view from a layout. It then generates a net list from each one and compares them. Nodes, ports, and device sizing are all compared. If they are the same, LVS passes and the designer can continue.

LVS tends to consider transistor fingers to be the same as an extra-wide transistor. For example, 4 transistors in parallel (each 1 μm wide), a 4-finger 1 μm transistor, and a 4 μm transistor are all seen as the same by the LVS tool. The functionality of .lib files will be taken from spice models and added as an attribute to the .lib file.

- **Ques: 141. What are the steps involved in semiconductor device fabrication?**

Solution:
This is a list of processing techniques that are employed numerous times in a modern electronic device and do not necessarily imply a specific order.

Wafer processing
Wet cleans
Photolithography
Ion implantation (in which dopants are embedded in the wafer creating regions of increased [or decreased] conductivity)
Dry etching
Wet etching
Plasma ashing
Thermal treatments
Rapid thermal anneal
Furnace anneals
Thermal oxidation
Chemical vapor deposition (CVD)
Physical vapor deposition (PVD)
Molecular beam epitaxy (MBE)
Electrochemical deposition (ECD) (see electroplating)
Chemical-mechanical planarization (CMP)
Wafer testing (where the electrical performance is verified)
Wafer back grinding (to reduce the thickness of the wafer so the resulting chip can be put into a thin device like a smartcard or PCMCIA card)
Die preparation
Wafer mounting
Die-cutting
IC packaging
Die attachment
IC Bonding
Wire bonding
Flip chip
Tab bonding
IC encapsulation
Baking
Plating
Laser marking
Trim and form
IC testing

- **Ques: 142. Name different types of the logic family.**

Solution:
Listed here in rough chronological order of introduction along with their usual abbreviations of logic family

Diode logic (DL)
Direct-coupled transistor logic (DCTL)
Complementary transistor logic (CTL)
Resistor-transistor logic (RTL)
Resistor-capacitor transistor logic (RCTL)
Diode-transistor logic (DTL)
Emitter coupled logic (ECL) also known as current-mode logic (CML)
Transistor–transistor logic (TTL) and variants
P-type metal–oxide–semiconductor logic (PMOS)
N-type metal–oxide–semiconductor logic (NMOS)
Complementary metal–oxide–semiconductor logic (CMOS)
Bipolar complementary metal–oxide–semiconductor logic (BiCMOS)
Integrated injection logic (I2L)

- **Ques: 143. What are the different types of IC packaging? Name any 10 types.**

Solution:
IC is packaged in many types as listed hereunder:

1. BGA1
2. BGA2
3. Ball grid array
4. CPGA
5. The ceramic ball grid array
6. DIP-8
7. Die attachment
8. Dual flat no-lead
9. Dual-in-line package
10. Flatpack
11. Land grid array
12. Leadless chip carrier
13. Low insertion force
14. MicroFCBGA
15. Multichip module
16. Pin grid array
17. Single in-line package
18. Surface-mount technology
19. Through-hole technology
20. Zig-zag in-line package

- **Ques: 144. What is substrate coupling?**

Solution:
In an IC, a signal can couple from one node to another via the substrate. This phenomenon is referred to as substrate coupling or substrate noise coupling.

The push for reduced cost, more compact circuit boards, and added customer features have provided incentives for the inclusion of analog functions on primarily digital MOS ICs forming mixed-signal ICs.

- **Ques: 145. What do you understand by "latchup?"**

Solution:
A latchup is the inadvertent creation of a low-impedance path between the power supply rails of an electronic component, triggering a parasitic structure, which then acts as a short circuit, disrupting the proper functioning of the part and possibly even leading to its destruction due to overcurrent. A power cycle is required to correct this situation. The parasitic structure is usually equivalent to a thyristor (or SCR), a PNPN structure that acts as a PNP and an NPN transistor stacked next to each other. During a latchup when one of the transistors is conducting, the other one begins conducting too. They both keep each other in saturation for as long as the structure is forward-biased and some current flows through it – which usually means until a power down. The SCR parasitic structure is formed as a part of the totem-pole PMOS and NMOS transistor pair on the output drivers of the gates.

- **Ques: 146. What is a crystal oscillator?**

Solution:
A crystal oscillator is an electronic oscillator circuit that uses the mechanical resonance of a vibrating crystal of piezoelectric material to create an electrical signal with a precise frequency.

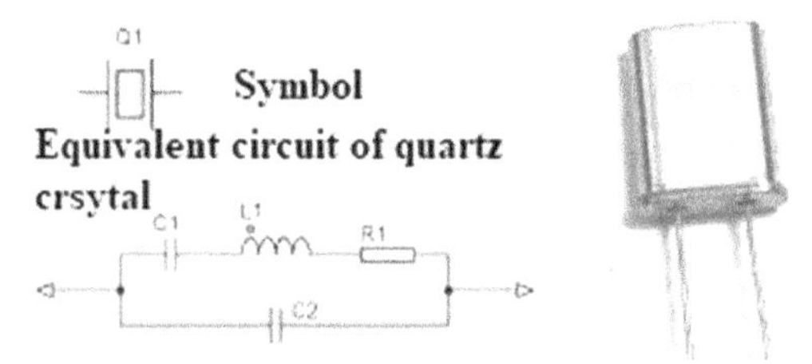

- **Ques: 147. Explain the concept of thyristor.**

Solution:
A thyristor is a solid-state semiconductor device with four layers of alternating P- and N-type materials. It acts exclusively as a bistable switch,

conducting when the gate receives a current trigger and continuing to conduct until the voltage across the device is reversed biased, or until the voltage is removed.

- **Ques: 148. What is the current-controlled and voltage-controlled device?**

Solution:
Current-controlled devices are those, whose output characteristic depends on the input current.

Voltage-controlled devices are those whose output depends on the input voltage.

- **Ques: 149. What do you mean by IGBT?**

Solution:
The insulated gate bipolar transistor (IGBT) is a semiconductor device with three terminals and is used mainly as an electronic switch. It is characterized by fast switching and high efficiency, which makes it a necessary component in modern appliances such as lamp ballasts, electric cars, and variable frequency drives (VFDs).

Its ability to turn on and off, rapidly, makes it applicable in amplifiers to process complex wave-patterns with pulse width modulation. IGBT combines the characteristics of MOSFETs and BJTs to attain high current and low saturation voltage capacity respectively. It integrates an isolated gate using FET (field effect transistor) to obtain a control input.

- **Ques:150. What is the principle of operation in IGBT?**

Solution:
IGBT requires only a small voltage to maintain conduction in the device unlike in BJT. The IGBT is a unidirectional device, that is, it can only switch ON in the forward direction. This means current flows from the collector to the emitter unlike in MOSFETs, which are bidirectional.

- **Ques: 151. What is cyclo-converter?**

Solution:
A cyclo-converter refers to a frequency changer that can change AC power from one frequency to AC power at another frequency. This process is known as AC-AC conversion. It is mainly used in electric traction, AC motors having variable speed and induction heating.

- **Ques: 152. What do you mean by linear circuit elements?**

Solution:
Linear circuit elements refer to the components in an electrical circuit that exhibit a linear relationship between the current input and the voltage output. Examples of elements with linear circuits include the following:

Resistors
Capacitors
Inductors
Transformers

- **Ques: 153. Write the symbol of various resistors.**

Solution:

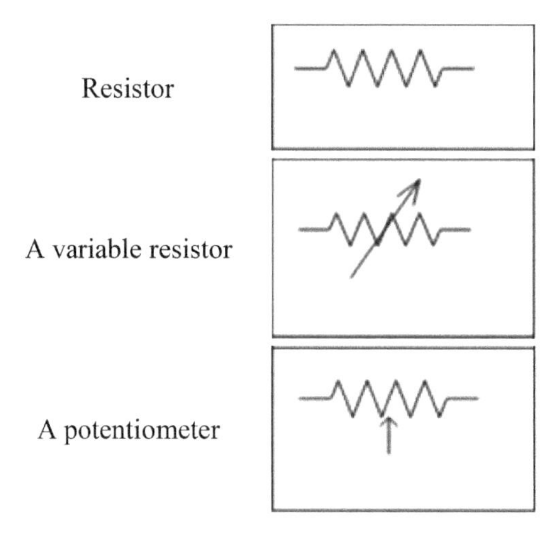

- **Ques: 154. Write the symbol of various capacitors.**

Solution:

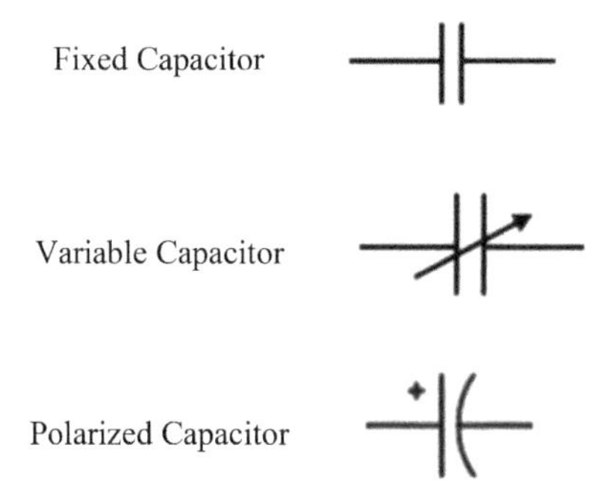

- **Ques: 155. Write the symbol of various inductors.**

Solution:

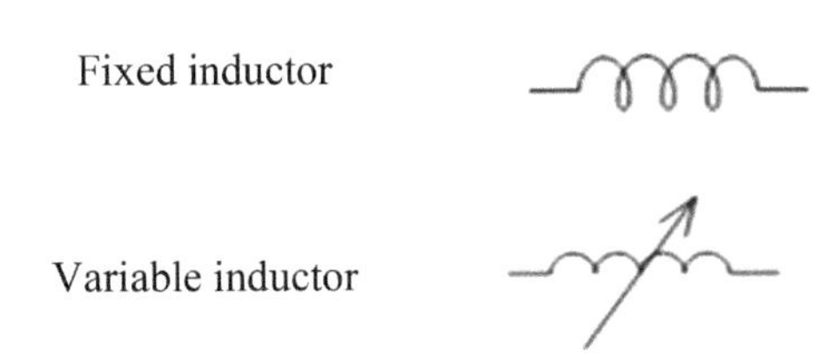

- **Ques: 156. What do you mean by the transformer?**

Solution:

This refers to a device that alters energy from one level to another through a process known as electromagnetic induction. It is usually used to raise or lower AC voltages in applications utilizing electric power.

When the current on the primary side of the transformer is varied, a varied magnetic flux is created on its core, which spreads out to the secondary windings of the transformer in the form of magnetic fields.

The operation principle of a transformer relies on Faraday's law of electromagnetic induction. The law states that the rate of change of the flux linking concerning time is directly related to the EMF induced in a conductor.

A transformer has three main parts as listed hereunder:

- ☐ Primary winding
- ☐ Magnetic core
- ☐ Secondary winding

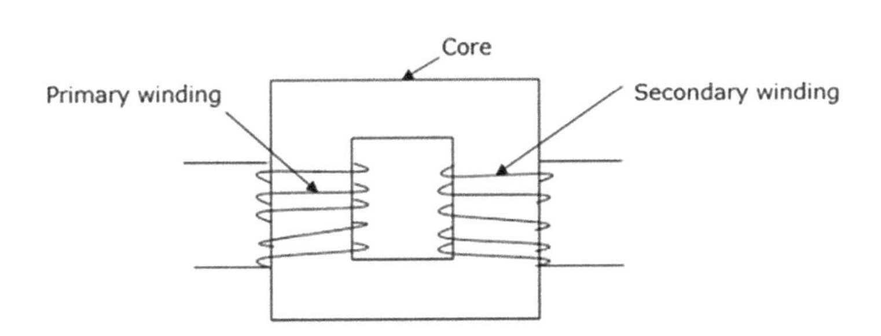

Symbol of a Transformer

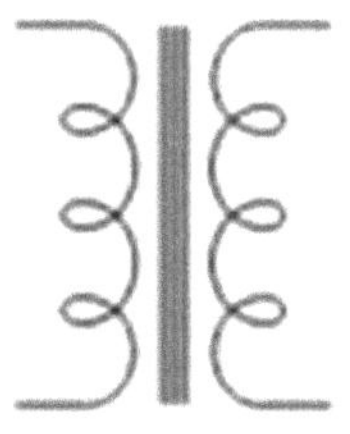

- **Ques: 157. What do you mean by solar cells?**

Solution:
A solar cell, or photovoltaic cell, is an electrical device that converts the energy of light directly into electricity by the photovoltaic effect, which is a physical and chemical phenomenon. Photovoltaic cells generate electricity by absorbing sunlight and using light energy to create an electrical current. There are many photovoltaic cells within a single solar panel, and the current created by all of the cells together adds up to enough electricity to help power your home.

- **Ques: 158. How many types of solar panels are there?**

Solution:
There are mainly two main types of solar panels:

1. Photovoltaic (PV) solar panels

The technology most people think of when they say "solar panels." These devices convert sunlight into electricity. For the sake of this chapter, the term "solar panels" will be used to describe photovoltaic panels.

2. Solar thermal collectors

They use the same solar energy that photovoltaic panels do, but they generate heat instead of electrical power.

Multiple solar cells that are oriented in the same way make up what we call solar panels. The electrical power out depends on how many of them are put together.

- **Ques: 159. What are solar cells? Explain its working.**

Solution:
Solar cells generate an electric current when they are exposed to light. Exactly how this happens is quite complex, and varies between the different types of solar panels.

The basic gist is this:

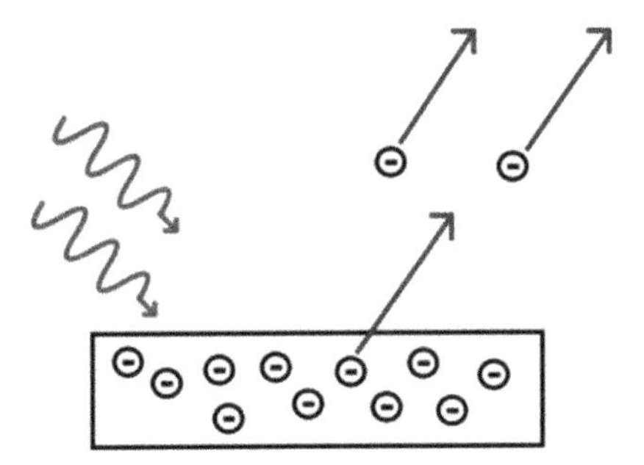

1. Incoming photons are absorbed by semiconducting material (in most cases silicon) on the surface of the solar cell.
2. These photons “knock loose” electrons from atoms in the solar cell. Since electrons carry a negative charge, an electric potential difference has been created.
3. The solar cell is built in a way that only allows the electron to move in one direction to cancel out the potential.
4. Put many of these reactions together and current starts flowing through the material.

- **Ques: 160. What do you understand by ‘LED’?**

Solution:
The lighting emitting diode is a p–n junction diode. It is a specially doped diode and made up of a special type of semiconductor. When the light emits in the forward biased, then it is called as an LED.

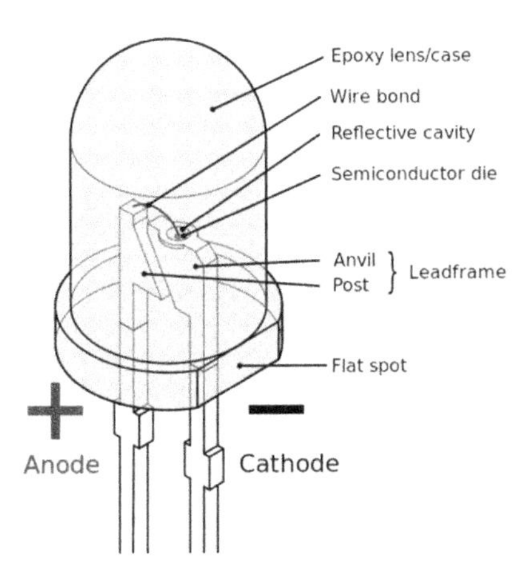

- **Ques: 161. What is the working of an LED?**

Solution:
The light-emitting diode simply, we know as a diode. When the diode is forward biased, then the electrons and holes are moving fast across the junction and they are combining constantly, removing one another out. Soon after the electrons are moving from the n-type to the p-type silicon, it combines with the holes, then it disappears. Hence it makes the complete atom and more stable and it gives the little burst of energy in the form of a tiny packet or photon of light.

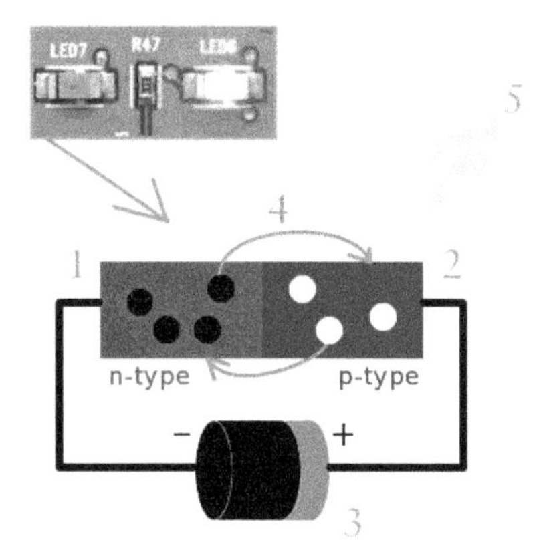

The above diagram shows how the LED works and the step-by-step process of the diagram.

- From the diagram, we can observe that the N-type silicon is in red and it contains the electrons, and they are indicated by the black circles.
- The P-type silicon is in blue and it contains holes, and they are indicated by the white circles.
- The power supply across the p–n junction makes the diode forward biased and pushing the electrons from n-type to p-type, thus pushing the holes in the opposite direction.
- Electron and holes at the junction are combined.
- The photons are given off as the electrons and holes are recombined.

- **Ques: 162. What is the working principle of an LED?**

Solution:
The working principle of the LED is based on quantum theory. The quantum theory says that when the electron comes down from the higher energy level

to the lower energy level then, the energy emits from the photon. The photon energy is equal to the energy gap between these two energy levels. If the P–N junction diode is forward biased, then the current flows through the diode.

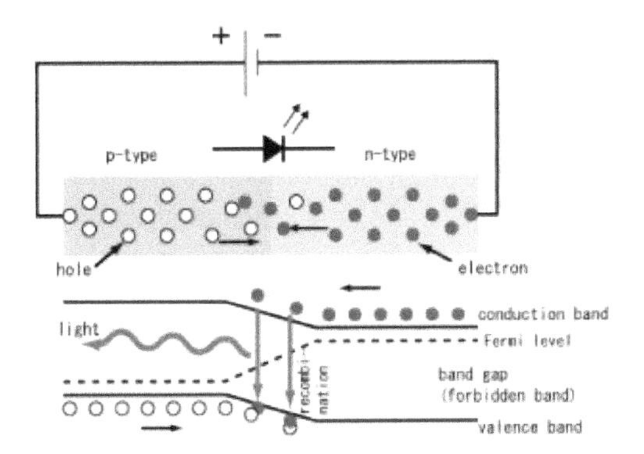

The flow of current in the semiconductors is caused by both the flow of holes in the opposite direction of the current and flow of electrons in the direction of the current. Hence, there will be recombination due to the flow of these charge carriers.

The recombination indicates that the electrons in the conduction band jump down to the valence band. When the electrons jump from one band to another band the electrons will emit the electromagnetic energy in the form of photons and the photon energy is equal to the forbidden energy gap.

For example, let us consider the quantum theory, the energy of the photon is the product of both Planck constant and frequency of electromagnetic radiation. The mathematical equation is shown as follows:

$$Eq = hf$$

where h is known as Planck constant, and the velocity of electromagnetic radiation is equal to the speed of light, i.e., c. The frequency radiation is related to the velocity of light as an $f = c / \lambda$. λ is denoted as a wavelength of electromagnetic radiation and the above equation will become

$$Eq = he/\lambda$$

From the above equation, we can say that the wavelength of electromagnetic radiation is inversely proportional to the forbidden gap. In general silicon, germanium semiconductors this forbidden energy gap is between the condition and valence bands are such that the total radiation of electromagnetic waves during recombination is in the form of the infrared radiation. We can't see the wavelength of infrared because they are out of our visible range.

The infrared radiation is said to be as heat because the silicon and the germanium semiconductors are not direct gap semiconductors rather these are indirect gap semiconductors. But in the direct gap semiconductors, the maximum energy level of the valence band and minimum energy level of the conduction band does not occur at the same moment of electrons. Therefore, during the recombination of electrons and holes are migration of electrons from the conduction band to valence band the momentum of the electron band will be changed.

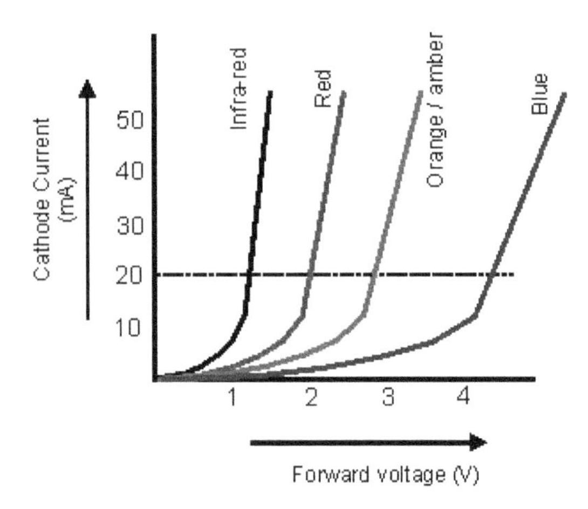

- **Ques: 163. What do you understand by "LDR?"**

Solution:
A light-dependent resistor (LDR) is also called a photoresistor or a cadmium sulfide (CdS) cell. An LDR is a component that has a (variable) resistance that changes with the light intensity that falls upon it. This allows them to be used in light-sensing circuits.

Variation in resistance with changing light intensity

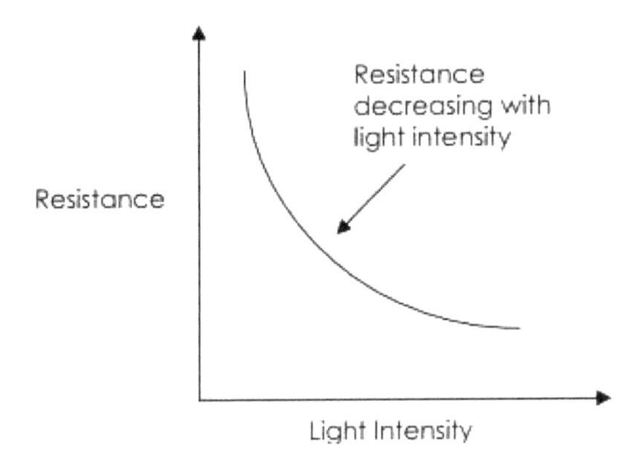

- **Ques: 164. What are the applications of LDR?**

Solution:
There are many applications for LDR. These include:

1. Lighting switch
   The most obvious application for an LDR is to automatically turn on a light at a certain light level. An example of this could be a street light or a garden light.
2. Camera shutter control
   LDRs can be used to control the shutter speed on a camera. The LDR would be used to measure the light intensity which then adjusts the camera shutter speed to the appropriate level.

- **Ques: 165. Why do we use silicon in solar cells?**

Solution:
Silicon is a semiconductor material. When it is doped with the impurities, gallium and arsenic, its ability to capture the sun's energy and convert it into electricity is improved considerably.

- **Ques: 166. What is "multivibrator?"**

Solution:
It is a device consisting of two amplifying transistors or valves, each with its output connected to the input of the other, which produces an oscillatory signal.

- **Ques: 167. What is the difference between amplifier and oscillators?**

Solution:
The amplifier is an electronic circuit that gives the output as an amplified form of input. The oscillator is an electronic circuit that gives output without the application of input. The amplifier does not generate any periodic signal.

- **Ques: 168. Why are CE amplifiers widely used?**

Solution:
CE is most widely used because it provides the voltage gain required for most of the day-to-day applications of preamp and power amps. The common emitter is the most basic configuration for amplifier circuits. It also provides the maximum transconductance or voltage gain for a given load.

- **Ques: 169. Why NPN is preferred over PNP?**

Solution:
Majority charge carriers in NPN are electrons, whereas in the case of PNP there are holes. Holes and electrons are charge carriers in a BJT (NPN or PNP). The difference between them is the mobility (with applied voltage), (effective) mass. Electrons are better when compared to holes so (NPN) is preferred.

- **Ques: 170. Why FET is named a field-effect transistor?**

Solution:
FET is named a field-effect transistor as field-effect is producing, but in BJT also field effect will be produced. The electric field is produced if there is a voltage difference. The field becomes nonzero when voltage differences are nonzero. FET works on the principle that it uses the field produced by the gate in a way that makes the channel conduct more or less.

- **Ques: 171. What is the cut-off frequency?**

Solution:
The frequency at which the response is –3 dB concerning the maximum response.

- **Ques: 172. Explain the concept of the rectifier.**

Solution:
A rectifier changes AC into DC. This process is called rectification. The three main types of rectifiers are half-wave, full-wave, and bridge. A rectifier is the opposite of an inverter, which changes DC into AC.

HWR – The simplest type is the half-wave rectifier, which can be made with just one diode. When the voltage of the AC is positive, the diode becomes forward-biased and current flows through it. When the voltage is negative, the diode is reverse-biased and the current stops.

The result is a clipped copy of the AC waveform with an only positive voltage, and an average voltage that is one-third of the peak input voltage. This pulsating DC is adequate for some components, but others require a steadier current. This requires a full-wave rectifier that can convert both parts of the cycle to positive voltage.

FWR: The full-wave rectifier is essentially two half-wave rectifiers and can be made with two diodes and an earthed center tap on the transformer. The positive voltage half of the cycle flows through one diode, and the negative half flows through the other. The center tap allows the circuit to be completed because the current cannot flow through the other diode. The result is still a pulsating DC but with just over half the input peak voltage, and double the frequency.

- **Ques: 173. What are transducer and transponder?**

Solution:
A transducer is a device, usually electrical, electronic, electromechanical, electromagnetic, photonic, or photovoltaic that converts one type of energy or physical attribute to another for various purposes including measurement or information transfer.

In telecommunication, the term transponder (short for transmitter-responder and sometimes abbreviated to XPDR, XPNDR, TPDR, or TP) has the following meanings:

- ☐ An automatic device that receives, amplifies, and retransmits a signal on a different frequency (see also broadcast translator).
- ☐ An automatic device that transmits a predetermined message in response to a predefined received signal.
- ☐ A receiver-transmitter that will generate a reply signal upon proper electronic interrogation.
- ☐ A communications satellite's channels are called transponders because each is a separate transceiver or repeater.

- **Ques: 174. What is an ideal voltage source?**

Solution:
It is a device with zero internal resistance.

- **Ques: 175. What is an ideal current source?**

Solution:
It is a device with infinite internal resistance.

- **Ques: 176. What is a practical voltage source?**

Solution:
It is a device with small internal resistance.

- **Ques: 177. What is a practical current source?**

Solution:
It is a device with large internal resistance.

- **Ques: 178. Design a flow chart depicting various types of capacitors?**

Solution:

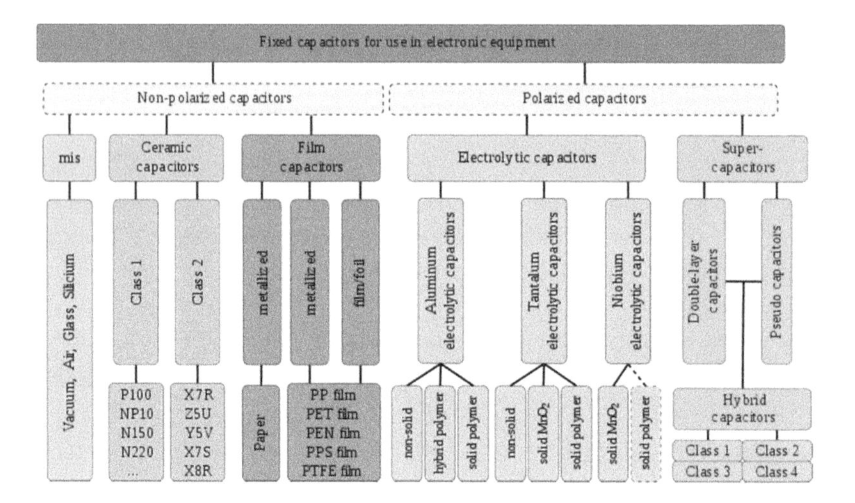

- **Ques:179. Show the various capacitors using photographs of components.**

Solution:
Some capacitors are manufactured so they can only tolerate applied voltage in one polarity but not the other. This is due to their construction: the dielectric is a microscopically thin layer of insulation deposited on one of the plates

by a DC voltage during manufacture. These are called electrolytic capacitors, and their polarity is marked.

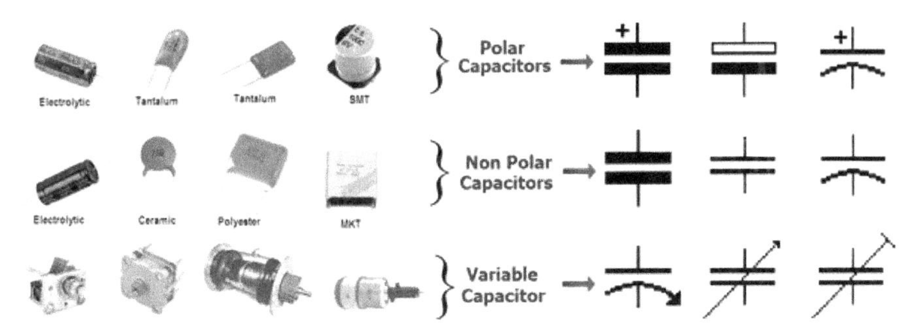

An electrolytic capacitor (abbreviated e-cap) is a polarized capacitor whose anode or positive plate is made of a metal that forms an insulating oxide layer through anodization. This oxide layer acts as the dielectric of the capacitor.

A polarized ("polar") capacitor is a type of capacitor that has an implicit polarity – it can only be connected one way in a circuit. The only reason people use polarized caps is that they often cost much less than nonpolarized caps of the same capacitance and voltage rating.

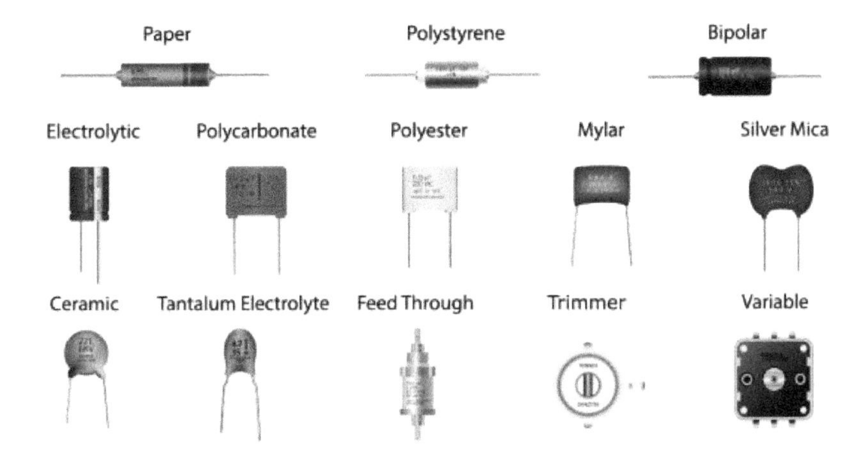

- **Ques: 180. Show the various resistors using photographs of components.**

Solution:

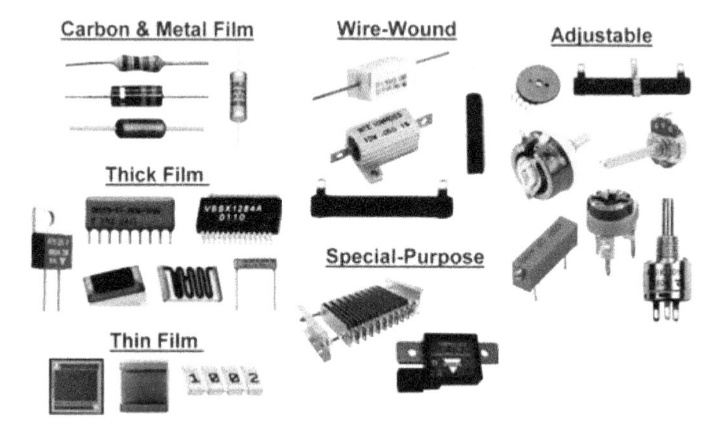

- Ques: 181. **Show the various sensors using photographs of components.**

Solution:

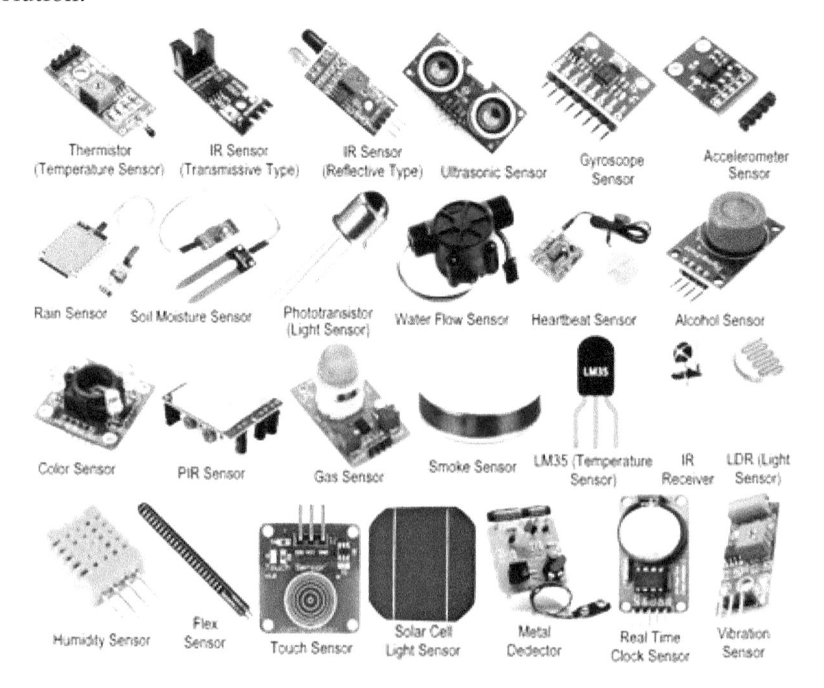

- **Ques: 182. Classify various types of solar cells.**

Solution:

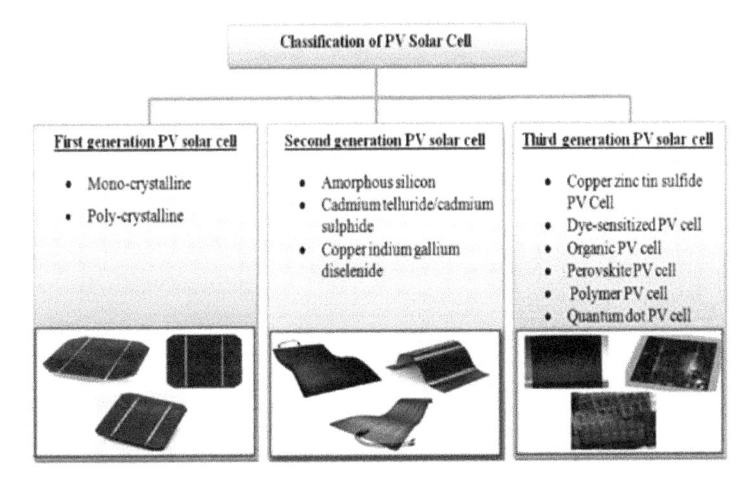

- **Ques: 183. What causes capacitors to leak?**

Solution:

There are two kinds of "leakage" in capacitors.

1. The gradual loss of charge in the capacitor. This occurs because the dielectric material is not a perfect insulator – it has a small but nonzero electrical conductivity, so it acts as a large resistance in parallel with the capacitor. Leakage can also occur through external components connected to the capacitor. This kind of leakage is why dynamic random-access memory (DRAM) must be constantly refreshed.

2. When the capacitor physically leaks its liquid electrolyte. This can occur in a very dramatic way if an electrolytic capacitor is connected backward, or if an internal short circuit develops due to a manufacturing defect. The high current through the short circuit generates gas, which causes pressure to build up. (I'm not sure if the gas is generated through evaporation or electrolysis of the electrolyte, or a combination of both.) This opens a safety vent (a part of the capacitor's casing designed to fail under high pressure), venting the electrolyte. The safety vent is supposed to prevent the pressure from building up to the point where the entire capacitor violently explodes; in cheap capacitors, sometimes it fails.

Capacitors can also leak electrolytes nonviolently, for example, due to corrosion of the case.

- **Ques: 184. Show the various inductors using photographs of components.**

Solution:

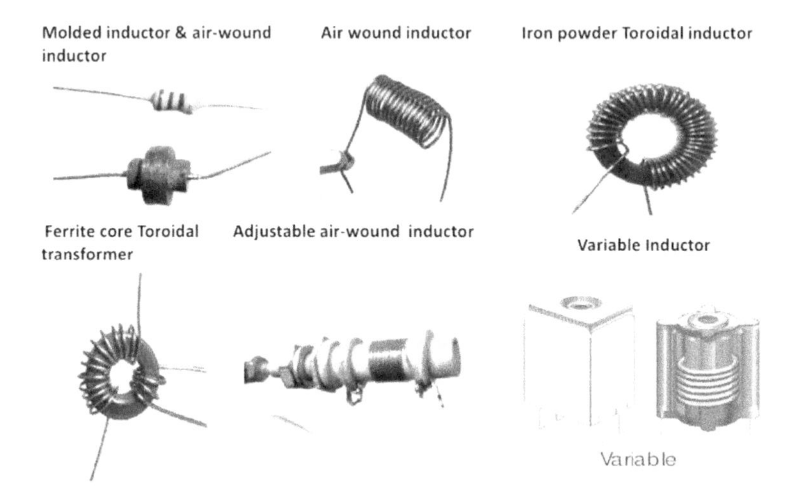

- **Ques: 185. Show the symbol of various types of inductors.**

Solution:

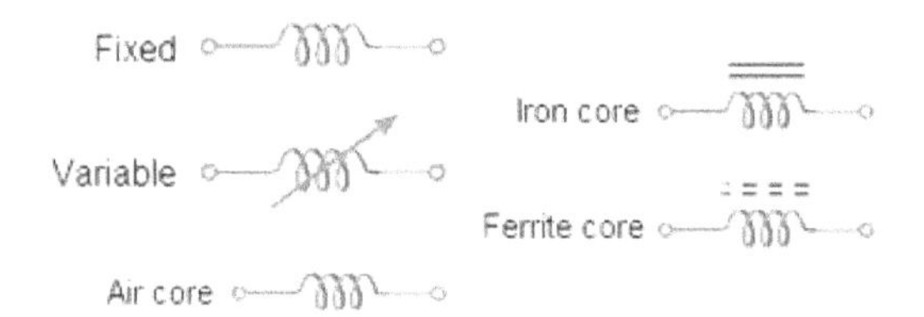

Series and Parallel Connections:

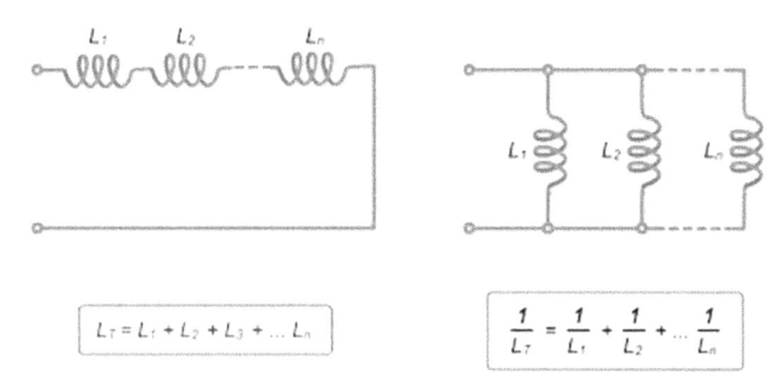

- **Ques: 186. Show the various optoelectronic devices using photographs.**

Solution:

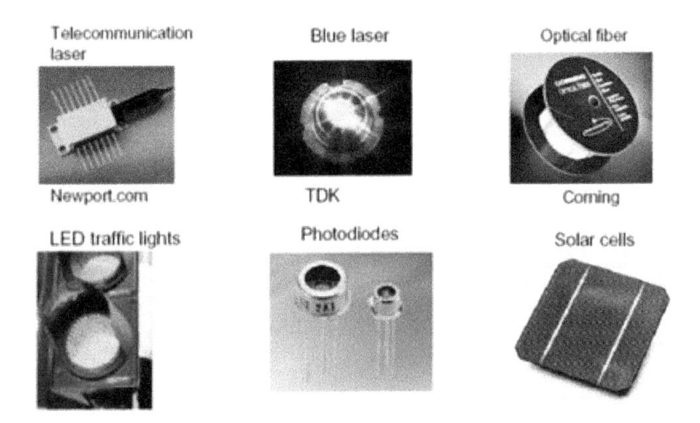

- **Ques: 187. Explain "Lenz law?"**

Solution:
When the current through a coil changes, a voltage is induced. Lenz's law states that the polarity of the induced voltage always opposes the change in current that caused it. The diagram above illustrates this law. When the switch closes, the current tries to increase, and the magnetic field starts expanding. The expanding magnetic field induces a voltage, which opposes any increase in current. So, at the instant of switching, the current remains the same. When the rate of expansion decreases, the induced voltage decreases, allowing the current to increase. As the current reaches a constant value, there is no induced voltage.

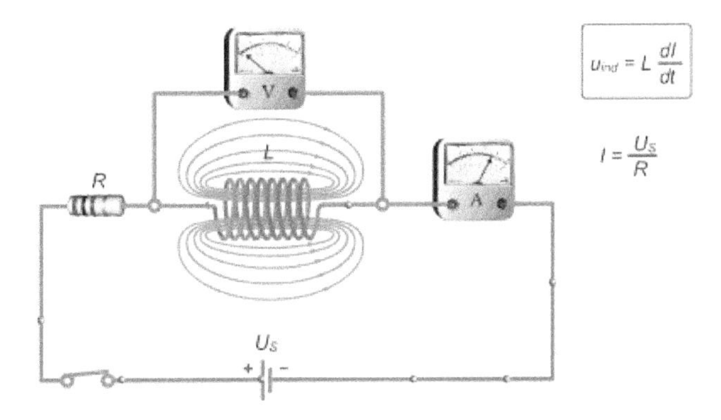

- **Ques: 188. What do you mean by "sine wave?"**

Solution:
The sine wave (sinusoidal wave or, simply, sinusoid) is the fundamental form of AC and voltage. The current reverses polarity over time. In one cycle, the polarity changes once. The time required for a given sine wave to complete one full cycle is called a period.

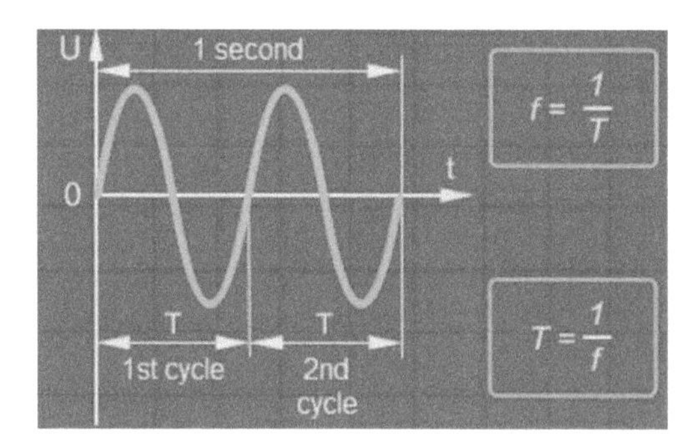

The number of cycles per second is the frequency (f), whose unit is the Hertz (Hz). One hertz is equal to one cycle per second. The frequency and period are reciprocal. More cycles per second results in a higher frequency and a shorter period.

- **Ques: 189. What is SOA for BJT?**

Solution:
The safe operating area (SOA) defines the current and voltage limitations of power devices.

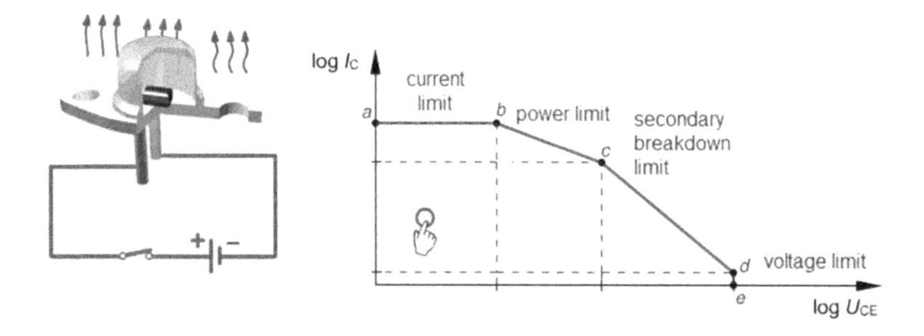

The above figure shows the typical SOA of a power bipolar transistor. It can be partitioned into four regions. The maximum current limit (sections a-b) and the maximum voltage limit (d-e) are determined by the technological features and construction of the particular device. The maximum power dissipation limits the product of the transistor's currents and voltages (sections b-c). A secondary breakdown (c-d) occurs when high voltages and high currents appear simultaneously when the device is turned off. When this happens, a hot spot is formed and the device fails due to thermal runaway.

- **Ques: 190. What is SOA for MOSFET and IGBT?**

Solution:

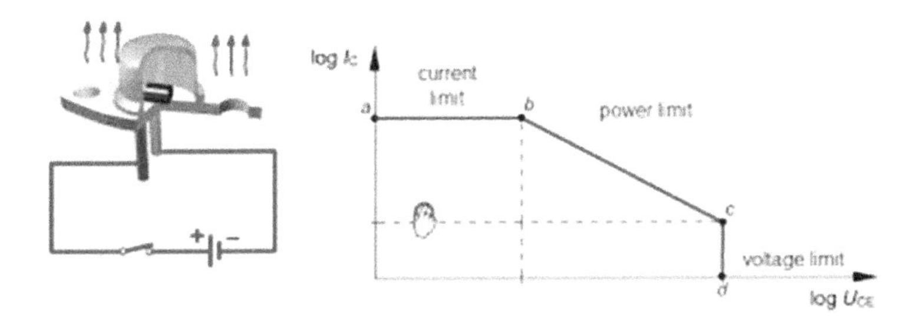

The above figure shows the safe operating area (SOA) of MOSFET transistors. This area is bounded by three limits: current limit (sections a-b), maximum power dissipation limit (b-c), and the voltage limits (c-d). The SOA of an IGBT is identical to that of the MOSFET SOA.

Since the drain current decreases when the temperature increases in MOSFET transistors, the possibility of secondary breakdown is almost nonexistent. If local heating occurs, the drain current – and consequently the power dissipation – both diminish. This avoids the creation of local hot spots that can cause thermal runaway.

The above figure demonstrates how the SOA of a device increases when the device is operating in pulse mode. When the device is operating in DC mode the safe operating area is at its smallest. The SOA grows when pulse mode is used. The shorter the pulse signal, the higher the SOA.

- **Ques: 191. What do you understand by maximum power dissipation?**

Solution:

The high currents and voltages in power devices produce very high internal power loss. This loss occurs in the form of heat that must be dissipated; otherwise, the device can be destroyed as a result of overheating.

The maximum power dissipation Pmax indicates a device's maximum capability to transfer and conduct this power loss without overheating.

- **Ques: 192. How maximum power dissipation and temperature are related?**

Solution:

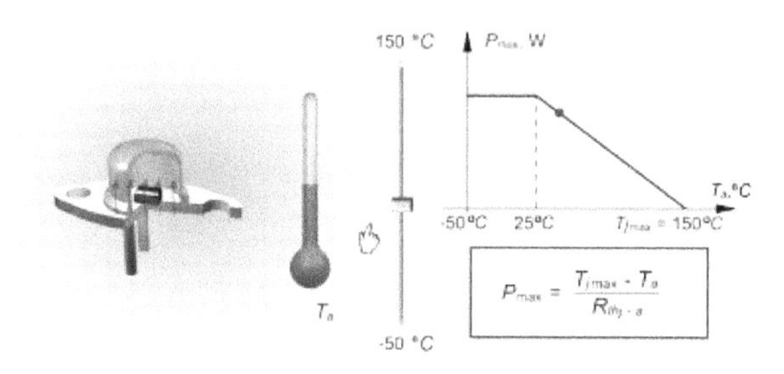

The maximum power dissipation Pmax of the transistor depends on the highest junction temperature that will not destroy the device Tjmax, the ambient temperature Ta, and the thermal resistance Rthj-a according to the equation shown in the above figure.

If the ambient temperature is less than or equal to 25°C the device reaches its maximum specified power rating. When the ambient temperature increases, the power rating decreases. If the ambient temperature Ta reaches the maximum junction temperature Tjmax, maximum power Pmax becomes zero.

- **Ques: 193. What do you mean by "heat sink?"**

Solution:

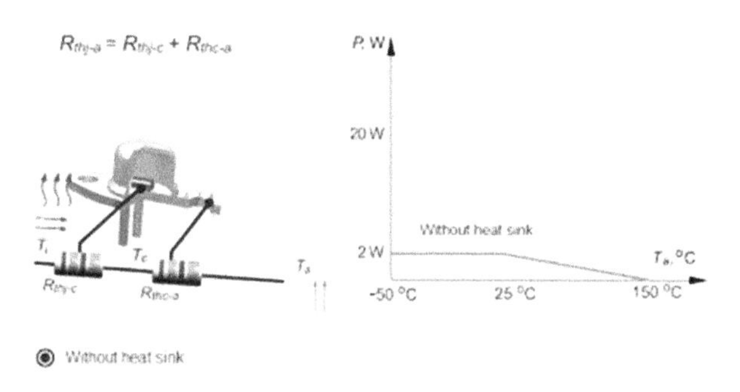

One way to increase the power rating of the device is to diminish the thermal resistance (Rthj-a). A heat sink, which is usually a metal construction with a large surface area, is used to allow heat to dissipate to ambient more easily.

When a heat sink is present, the global thermal resistance (Rthj-a) decreases because there are more paths available for heat dissipation. The case-to-heat-sink thermal resistance (Rthc-s) and the heat-sink-to-ambient thermal resistance (Rths-a) both facilitate heat dissipation. As a result, the power rating increases as illustrated in the above figure.

- **Ques: 194. What is the power transistor?**

Solution:
Power transistors are electronic components that are used for the control and regulation of voltages and currents with high values. They are the basic components for the implementation of linear and switched-mode power supplies, motor control circuits, automotive and aerospace systems, home appliances, and energy management systems.

- **Ques: 195. Why power transistors are necessary?**

Solution:
Power transistors are used to produce, convert, control, and regulate high amounts of power output. Typical headphone amplifiers have a low output value (just a fraction of a watt). They are usually implemented with standard low power transistors. On the other hand, amplifiers with a 100-watt output power are used to ensure quality sound in large rooms or concert halls. These amplifiers operate with high-level currents and voltages (more than dozens of amperes and volts). The output stages of such amplifiers can be implemented only with power transistors.

Power transistors are capable of providing high currents and high blocking voltages and therefore, high power. They can be classified into BJT (bipolar junction transistors), MOSFET (metal–oxide–semiconductor field effect transistors), and devices such as IGBT (insulated gate bipolar transistors) that combine bipolar and MOS technologies.

The principle of operation behind high power transistors is conceptually the same as bipolar or MOS transistors. The main difference is that the active area of the power devices is distinctly higher, resulting in a much higher current handling capacity. For this reason, they have large packages.

- **Ques: 196. Draw NPN BJT, N-channel MOSFET, and IGBT device.**

Solution:
The following figure depicts the typical structure of BJT, MOSFET, and IGBT devices. For a BJT to maintain conduction, a high continuous current through the base region is required. This imposes the necessity of high-power drive circuits.

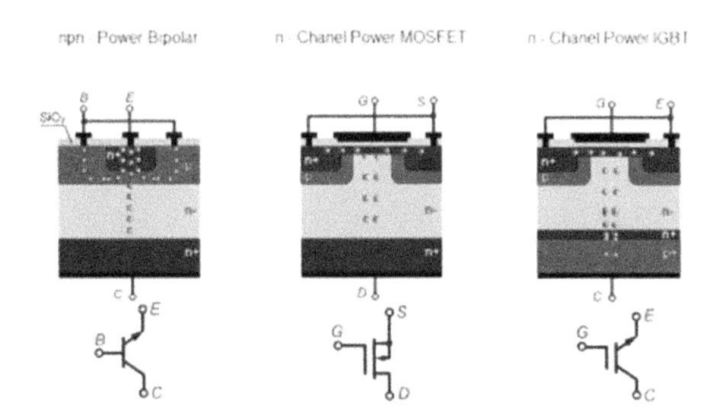

MOSFETs and IGBTs are voltage-controlled devices. The IGBT has one more junction than the MOSFET, which allows for a higher blocking voltage but limits the switching frequency. In IGBTs, during conduction, the holes from the collector p+ region are injected into the n-region. The accumulated charge reduces IGBT's on-resistance and thus the collector-to-emitter voltage drop is also reduced.

- **Ques: 197. What is the instantaneous and peak value of the sine wave?**

Solution:

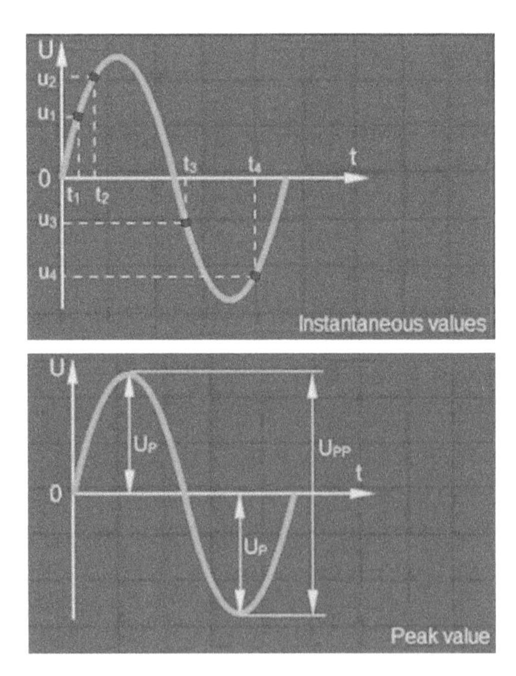

At any point in time on a sine wave, the voltage has an instantaneous value. As a cycle represents a continuous set of instantaneous values, other dimensions have been defined to enable comparing one wave to another. The peak value Up is the maximum value. It applies to either the positive or negative peak. The peak-to-peak value,Upp, is the voltage (or current) from the positive peak to the negative peak. The average value is an arithmetic average of all the values in a sine wave for one half-cycle, where Uavr = 0.637 Up.

- **Ques: 198. What do you understand by the "RMS value" of sine wave?**

Solution:

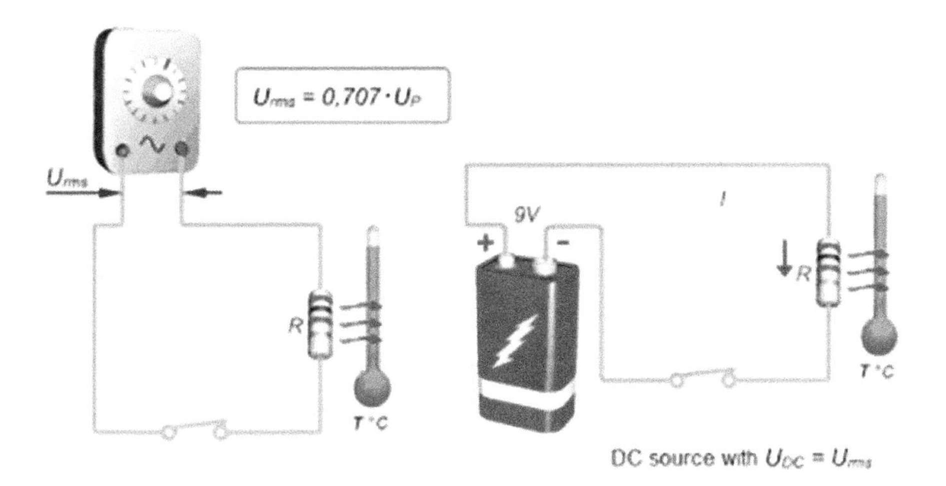

To compare AC and DC voltage, the effective value of the AC voltage should be calculated using the root mean square (RMS) value of the sinusoidal voltage. The RMS value of a sinusoidal voltage or current is equal to the DC voltage and current that produces the same heating effect. The formula is Urms = 0.707 Up. The factor 0.707 for RMS value is derived as the square root of the average (mean) of all the squares of the sine wave. To convert from RMS to peak value, the formula Up = 1.414 Urms is used. Unless indicated otherwise, all sine wave AC measurements are in RMS values.

- **Ques: 199. What is the phase angle of the sine wave?**

Solution:

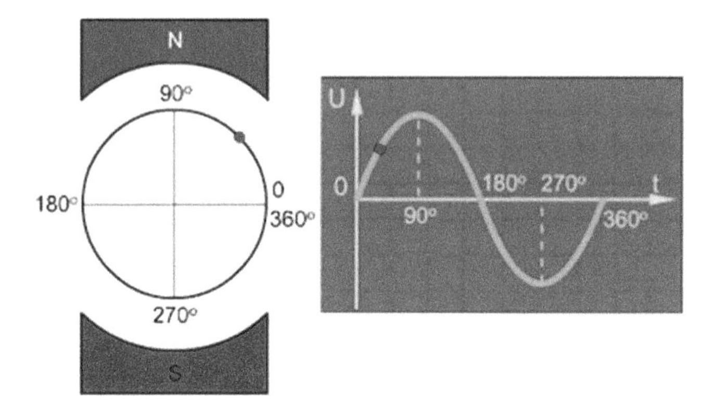

The angular measurement of a sine wave can be related to the angular rotation of an AC generator, as shown in the diagram above. It is based on 360° rotation for the complete cycle of a sine wave. The diagram shows angles in degrees over the full cycle of a sine wave. Since 360° = 2π rad, angles can be also expressed in radians using the formula in the illustration above.

The phase angle of a sine wave specifies the position of that sine wave relative to a reference. The illustration shows the phase shifts of a sine wave. There is a phase angle of 30° between sine wave A and sine wave B.

- **Ques: 200. What are the laws of resistive AC circuits?**

Solution:

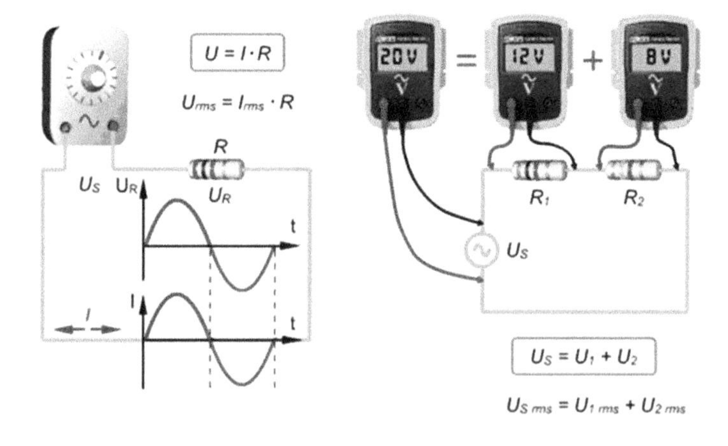

Ohm's law and Kirchoff's law apply to AC circuits in the same way they apply to DC circuits. If a sinusoidal voltage is applied across a resistor, there is a sinusoidal current. It is zero when the voltage is zero and is max when the voltage is max. The voltage and the currents are in phase with each other.

In a resistive circuit that has an AC voltage source, the source voltage is the sum of all the voltage drops, just as in a DC circuit. Remember, both the voltage and the current must be expressed in the same way, i.e., both in RMS, both in peak, etc.

- **Ques: 201. What are nonsinusoidal AC waveforms?**

Solution:

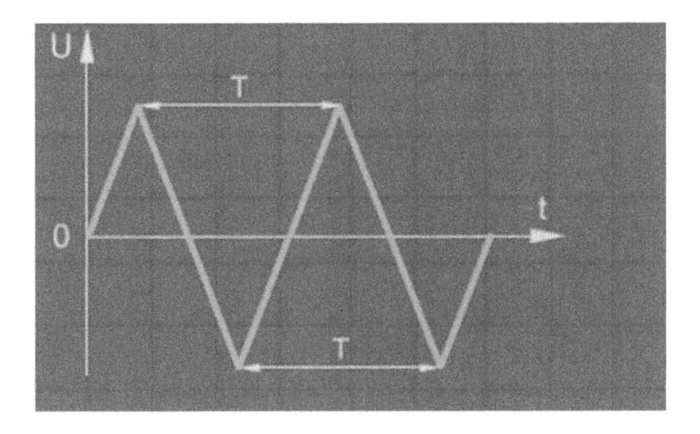

The pulse and the triangular waveform are the other two major types of signals widely used in electronics. Any waveform that repeats itself at fixed intervals is periodic. The period is denoted with a T. Triangular waveform are formed by voltage or current ramps. A ramp is a linear increase or decrease in the voltage.

An ideal pulse consists of two equal but opposite steps separated by an interval of time called the pulse width. The duty cycle of the pulse is the ratio of the pulse width to the period and is usually expressed in a percentage.

- **Ques: 202. What do you know about breadboard?**

Solution:
A breadboard is a solderless device for a temporary prototype with electronics and test circuit designs. Most electronic components in electronic circuits can be interconnected by inserting their leads or terminals into the holes and then making connections through wires where appropriate.

- **Ques: 203. How series and parallel connections are done using breadboard?**

Solution:
Series connection:

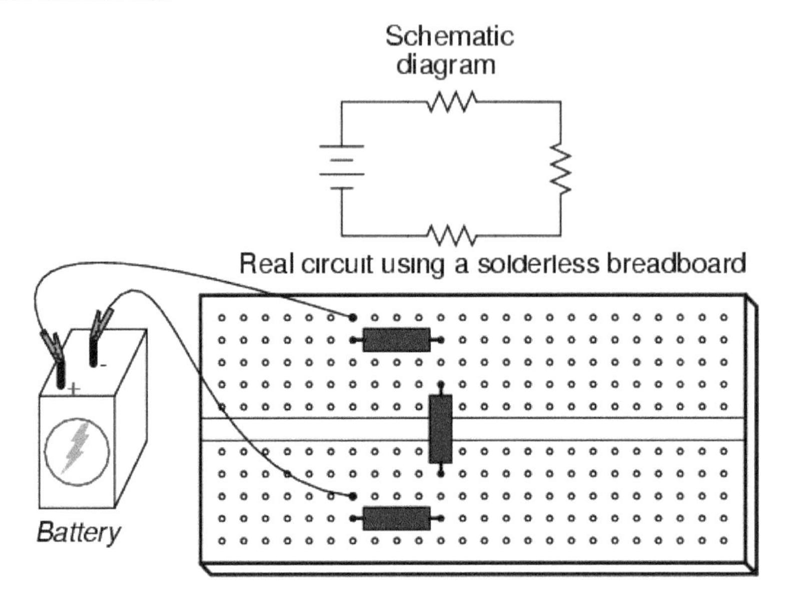

Parallel connection:

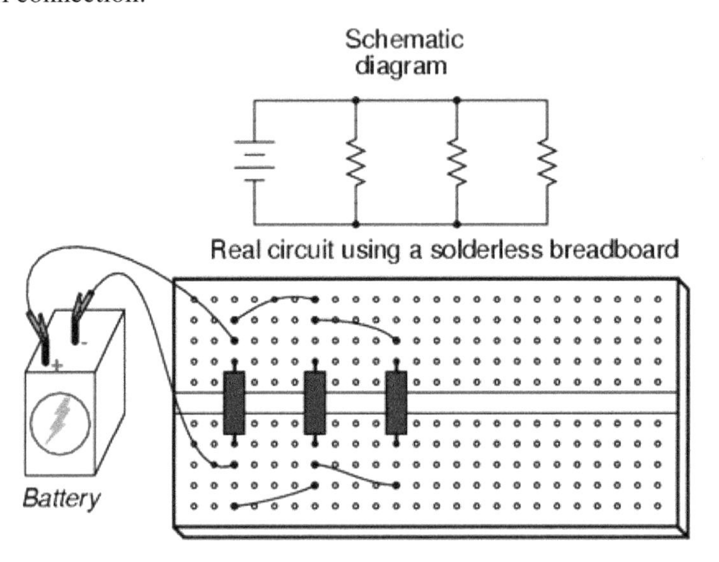

- **Ques: 204. What do you mean by worst-case analysis?**

Solution:
Worst-case circuit analysis (WCCA or WCA) is a cost-effective means of screening design to ensure with a high degree of confidence that potential defects and deficiencies are identified and eliminated before and during the test, production, and delivery.

It is a quantitative assessment of the equipment performance, accounting for manufacturing, environmental and aging effects. In addition to circuit analysis, a WCCA often includes stress and derating analysis, failure modes and effects criticality (FMECA) and reliability prediction (MTBF).

- **Ques: 205. What is the need for design verification and reliability?**

Solution:

1. Verifies circuit operation and quantifies the operating margins over part tolerances and operating conditions
2. Improve circuit performance – Determines the sensitivity of components to certain characteristics or tolerances to better optimize/understand design and what drives performance
3. Verifies that a circuit interfaces with another design properly
4. Determines the impact of part failures or out of tolerance modes

- **Ques: 206. What is a catastrophic failure?**

Solution:
A catastrophic failure is a sudden and total failure from which recovery is impossible. Catastrophic failures often lead to cascading systems failure.

- **Ques: 207. How sine waves are produced?**

Solution:
Sine waves are produced electromagnetically by an AC generator or electronically by an oscillator circuit, which is used in a signal generator.

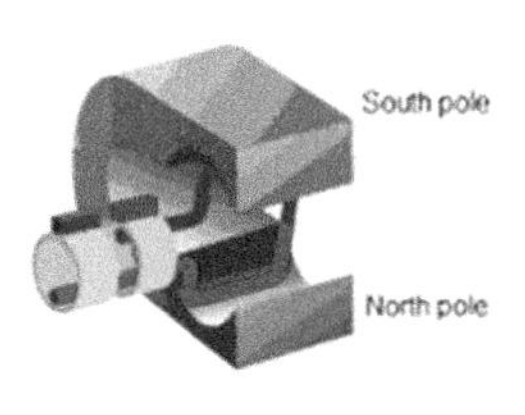

AC Generator

Simplified Model

The figure shows a cross-section of an AC generator. A simplified model of this generator consists of a single loop of wire in a permanent magnetic field. Magnetic flux lines exist around the north and south poles of the magnet. When a conductor rotates through the magnetic field, a voltage is induced.

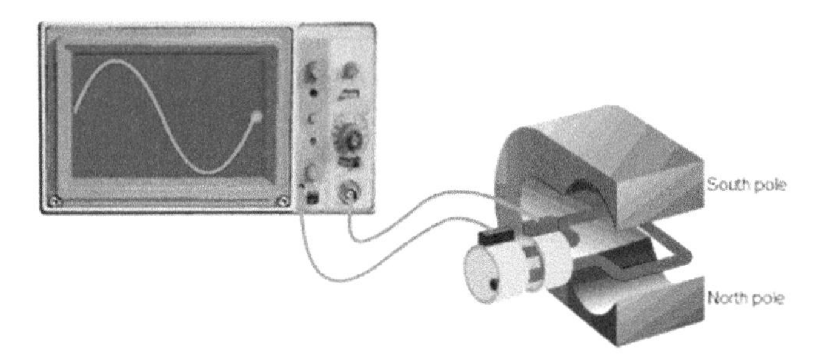

In a horizontal starting position, the loop does not induce a voltage because the conductors are not cutting across the magnetic flux lines. As the loop rotates through the first quarter of the cycle, it cuts through the flux lines producing the maximum induced voltage. During the second quarter of the cycle, the voltage decreases from its positive maximum back to zero. During the second half of the revolution, the wire loop cuts through the magnetic field in the opposite direction. Thus, the induced voltage has the opposite polarity. After one complete revolution of the loop, one full cycle of the sinusoidal voltage has been completed.

- **Ques: 208. What is a signal generator and cathode ray oscilloscope?**

Solution:
A signal generator is an instrument that electronically produces sinusoidal voltages or other types of waveforms whose amplitude and frequency can be adjusted. A typical signal generator is shown in the illustration.

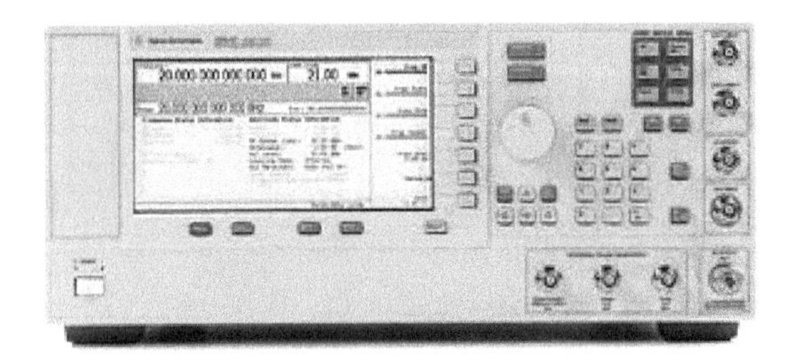

An oscilloscope previously called an oscillograph, and informally known as a scope or o-scope, CRO, or DSO is a type of electronic test instrument that graphically displays varying signal voltages, usually as a two-dimensional plot of one or more signals as a function of time.

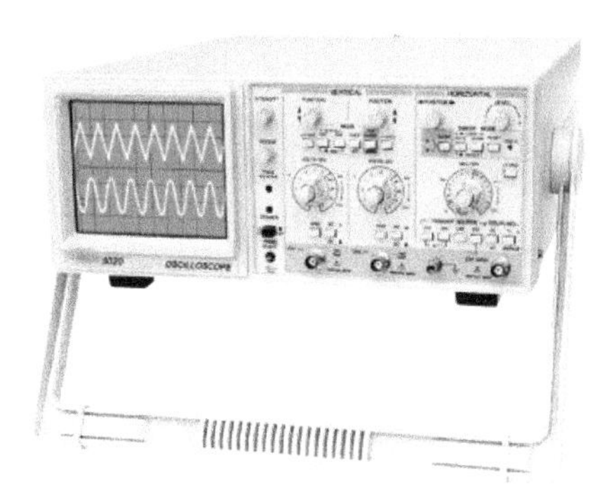

The cathode ray oscilloscope is an electronic test instrument, it is used to obtain waveforms when the different input signals are given.

- **Ques: 209. What is the difference between the bypass capacitor and the decoupling capacitor?**

Solution:
The coupling capacitor is used to maintain the DC biasing condition of circuit (different amplifiers, etc) cannot pass DC through it so that DC voltage at different terminals remains the same. However, the by-pass capacitor is used to pass the AC which passes through the resistor so that the effective gain will be higher for the same circuit.

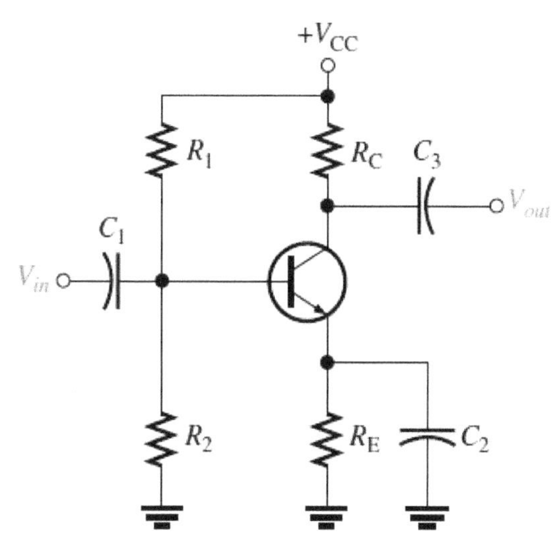

Here C1 and C3 are coupling capacitors and C2 is a by-pass capacitor.

The terms "bypass capacitor" and "decoupling capacitor" are used interchangeably, though there are definite differences between them.

The by-pass capacitor ("by-pass") helps us meet this requirement by constraining the unwanted communications aka the "noise" emanating from the power line to the electronic circuit in question. Any glitch or noise appearing on the power line is immediately bypassed into the chassis ground ("GND") and thus prevented from entering into the system, hence the name by-pass capacitor.

Decoupling capacitors ("decap"), on the other hand, are used to isolate two stages of a circuit so that these two stages don't have any DC effect on each other. In reality, decoupling is a refined version of bypassing.

- **Ques: 210. What do you mean by SOC?**

Solution:
As the name suggests, it means shrinking the whole system onto a single chip. The most important feature of the chip is that its functionality should be comparable to that of the original system. It improves quality, productivity, and performance.

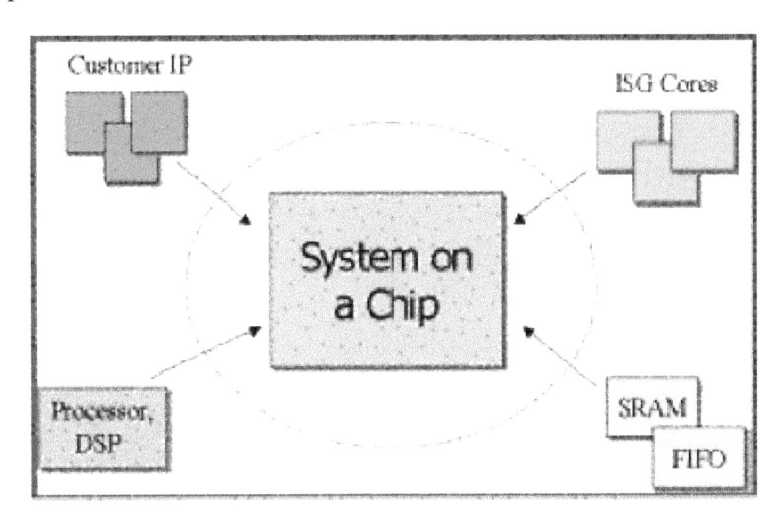

- **Ques: 211. What do you mean by a diode?**

Solution:
A diode is defined as a two-terminal electronic component that only conducts current in one direction. A diode will have negligible resistance in one

direction (to allow current flow), and very high resistance in the reverse direction (to prevent current flow).

A P–N junction is the simplest form of the semiconductor diode. In ideal conditions, this P–N junction behaves as a short circuit when it is forward biassed, and as an open circuit when it is in the reverse biased. The name diode is derived from "di-ode" which means a device that has two electrodes.

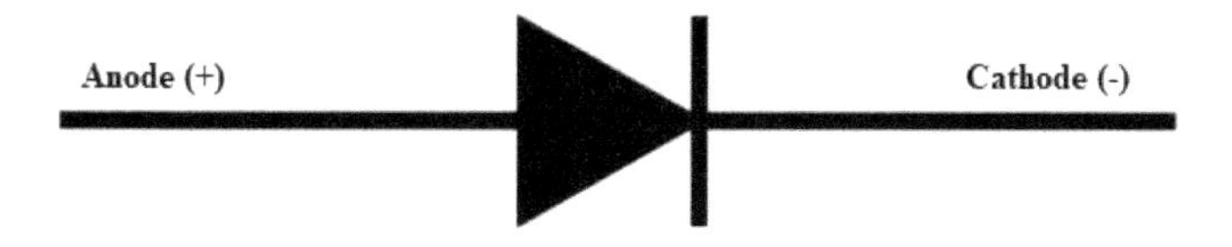

The arrowhead points in the direction of conventional current flow in the forward-biased condition. That means the anode is connected to the p-side and the cathode is connected to the n-side.

- **Ques: 212. What are the main applications of diode?**

Solution:
Diodes can be used as rectifiers, signal limiters, voltage regulators, switches, signal modulators, signal mixers, signal demodulators, and oscillators.

- **Ques: 213. What is the basic principle of the diode?**

Solution:
Semiconductor diodes are the most common type of diode. These diodes begin conducting electricity only if a certain threshold voltage is present in the forward direction (i.e., the "low resistance" direction). The diode is said to be "forward biased" when conducting current in this direction. When connected within a circuit in the reverse direction (i.e., the "high resistance" direction), the diode is said to be "reverse biased."

- **Ques: 214. Why do we say that diodes have a high resistance in the reverse direction, not an infinite resistance?**

Solution:
A diode only blocks current in the reverse direction (i.e., when it is reverse biased) while the reverse voltage is within a specified range. Above this range, the reverse barrier breaks. The voltage at which this breakdown occurs is called the "reverse breakdown voltage." When the voltage of the circuit is higher than the reverse breakdown voltage, the diode can conduct electricity in the reverse direction (i.e., the "high resistance" direction). This is the reason that diodes have a high resistance in the reverse direction, not an infinite resistance.

- **Ques: 215. Give the working principle of P–N junction diode in forward bias?**

Solution:
In forward bias, the positive terminal of a source is connected to the p-type side and the negative terminal of the source is connected to the n-type side of the diode and if we increase the voltage of this source slowly from zero, the diode will be in forwarding-biased state.

In the beginning, there is no current flowing through the diode. This is because although there is an external electrical field applied across the diode still the majority charge carriers do not get sufficient influence of the external field to cross the depletion region. We know that the depletion region acts as a potential barrier against the majority charge carriers. This potential barrier is called forward potential barrier. The majority charge carriers start crossing the forward potential barrier only when the value of externally applied voltage across the junction is more than the potential of the forward barrier. For silicon diodes, the forward barrier potential is 0.7 volt and for germanium diodes, it is 0.3 volt.

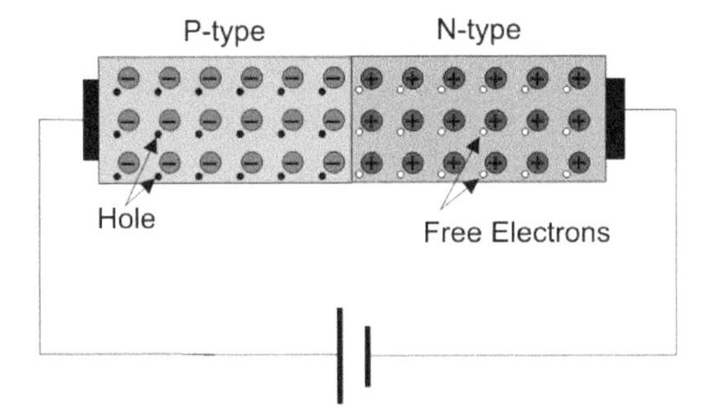

When the externally applied forward voltage across the diode becomes more than the forward barrier potential, the free majority charge carriers start crossing the barrier and contribute the forward diode current. In that situation, the diode would behave as a short-circuited path and the forward current gets limited by only externally connected resistors to the diode.

- **Ques: 216. Give the working principle of P–N junction diode in reverse bias.**

Solution:
A diode is said to be operated in reverse bias condition if we connect the negative terminal of the voltage source to the p-type side and positive terminal

of the voltage source to the n-type side of the diode. In that condition, due to electrostatic attraction of the negative potential of the source, the holes in the p-type region would be shifted more away from the junction leaving more uncovered negative ions at the junction. In the same way, the free electrons in the n-type region would be shifted more away from the junction toward the positive terminal of the voltage source leaving more uncovered positive ions in the junction. As a result of this phenomenon, the depletion region becomes wider. This condition of a diode is called the reverse-biased condition. In that condition, no majority carriers across the junction as they go away from the junction. In this way, a diode blocks the flow of current when it is reverse biased.

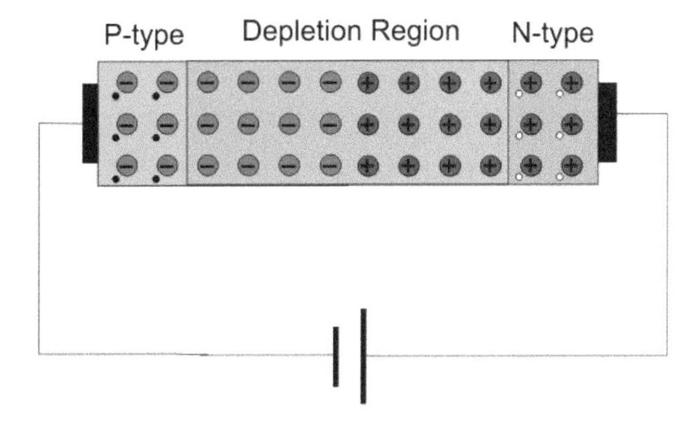

There are always some free electrons in the p-type semiconductor and some holes in the n-type semiconductor. These opposite charge carriers in a semiconductor are called minority charge carriers. In the reverse-biased condition, the holes find themselves in the n-type side would easily cross the reverse-biased depletion region as the field across the depletion region does not present rather it helps minority charge carriers to cross the depletion region. As a result, there is a tiny current flowing through the diode from positive to the negative side. The amplitude of this current is very small as the number of minority charge carriers in the diode is very small. This current is called reverse saturation current.

If the reverse voltage across a diode gets increased beyond a safe value, due to higher electrostatic force and due to higher kinetic energy of minority charge carriers colliding with atoms, several covalent bonds get broken to contribute a huge number of free electron–hole pairs in the diode and the process is cumulative. The huge number of such generated charge carriers would contribute a huge reverse current in the diode. If this current is not limited by an external resistance connected to the diode circuit, the diode may permanently be destroyed.

- **Ques: 217. Name the various types of diodes.**

Solution:

- Zener diode
- P–N junction diode
- Tunnel diode
- Varactor diode
- Schottky diode
- Photodiode
- PIN diode
- Laser diode
- Avalanche diode
- LED

- **Ques: 218. Explain the concept of Zener diode in brief.**

Solution:
Zener diode is a p–n junction diode connected in reverse bias. But ordinary P–N junction diode connected in reverse biased condition is not used as Zener diode practically. A Zener diode is a specially designed, highly doped P–N junction diode.

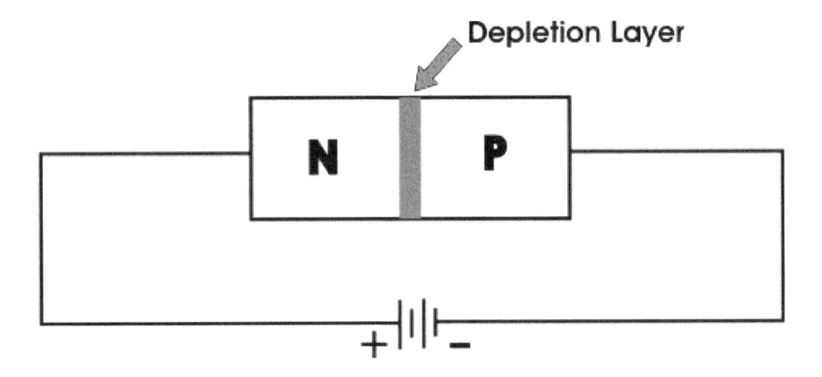

When a P–N junction diode is reverse biased, the depletion layer becomes wider. If this reverse-biased voltage across the diode is increased continually, the depletion layer becomes more and wider. At the same time, there will be a constant reverse saturation current due to minority carriers.

After a certain reverse voltage across the junction, the minority carriers get sufficient kinetic energy due to the strong electric field. Free electrons with sufficient kinetic energy collide with stationary ions of the depletion layer and knock out more free electrons. These newly created free electrons also get sufficient kinetic energy due to the same electric field, and they create more free electrons by collision cumulatively. Due to this commutative phenomenon, very soon, huge free electrons get created in the depletion layer, and the entire diode will become conductive. This type of breakdown of the depletion layer is known as an avalanche breakdown, but this breakdown is not quite sharp.

There is another type of breakdown in the depletion layer which is sharper compared to avalanche breakdown, and this is called Zener breakdown. When a P–N junction is diode is highly doped, the concentration of impurity atoms will be high in the crystal. This higher concentration of impurity atoms causes a higher concentration of ions in the depletion layer hence for the same applied reverse-biased voltage, the width of the depletion layer becomes thinner than that in a normally doped diode.

Due to this thinner depletion layer, voltage gradient or electric field strength across the depletion layer is quite high. If the reverse voltage is continued to increase, after a certain applied voltage, the electrons from the covalent bonds within the depletion region come out and make the depletion region conductive. This breakdown is called Zener breakdown. The voltage at which this breakdown occurs is called Zener voltage.

If the applied reverse voltage across the diode is more than Zener voltage, the diode provides a conductive path to the current through it hence, there is no chance of further avalanche breakdown in it. Theoretically, Zener breakdown occurs at a lower voltage level then avalanche breakdown in a diode, especially doped for Zener breakdown. The Zener breakdown is much sharper than the avalanche breakdown. The Zener voltage of the diode gets adjusted during manufacturing with the help of required and proper doping. When a Zener diode is connected across a voltage source, and the source voltage is more than Zener voltage, the voltage across a Zener diode remains fixed irrespective of the source voltage. Although at that condition current through the diode can be of any value depending on the load connected with the diode. That is why we use a Zener diode mainly for controlling voltage in different circuits.

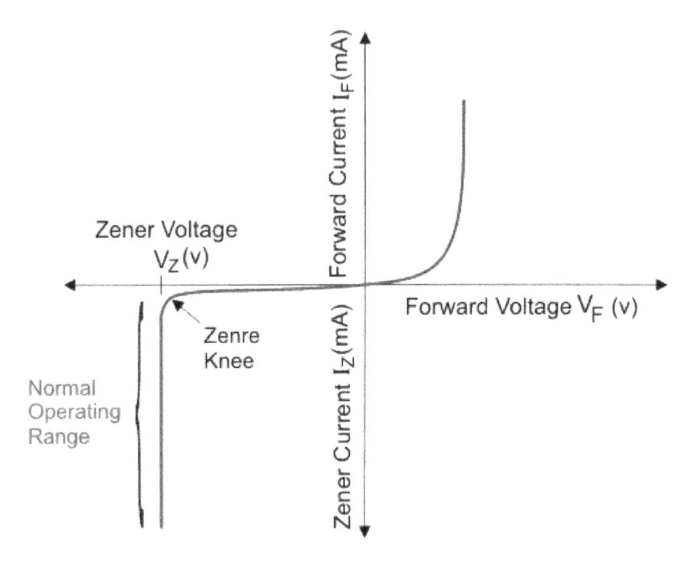

The above diagram shows the V–I characteristics of a Zener diode. When the diode is connected in forward bias, this diode acts as a normal diode but when the reverse bias voltage is greater than Zener voltage, a sharp breakdown takes place. In the V–I characteristics above V_z is the Zener voltage. It is also the knee voltage because at this point the current increases very rapidly.

- **Ques: 219. Explain P–N diode characteristics equation.**

Solution:
Let us consider a P–N junction with a donor concentration ND and acceptor concentration NA. Let us also assume that all the donor atoms have donated free electrons and become positive donor ions and all the acceptor atoms have accepted electrons and created corresponding holes and become negative acceptor ions. So, we can say the concentration of free electrons (n) and donor ions ND are the same and similarly, the concentration of holes (p) and acceptor ions (NA) are the same. Here, we have ignored the holes and free electrons created in the semiconductors due to unintentional impurities and defects.

$$n = N_D \quad and \quad p = N_A$$

Across the P–N junction, the free electrons donated by donor atoms in n-type side diffuse to the p-type side and recombine with holes. Similarly, the holes created by acceptor atoms in the p-type side diffuse to the n-type side and recombine with free electrons. After this recombination process, there is a lack of or depletion of charge carriers (free electrons and holes) across the junction. The region across the junction where the free charge carriers get depleted is called the depletion region. Due to the absence of free charge carriers (free electrons and holes), the donor ions of n-type side and acceptor ions of the p-type side across the junction become uncovered. These positive uncovered donor ions toward the n-type side adjacent to the junction and negative uncovered acceptors ions toward the p-type side adjacent to the junction cause a space charge across the P–N junction. The potential developed across the junction due to this space charge is called the diffusion voltage. The diffusion voltage across a P–N junction diode can be expressed as

$$V_D = \frac{kT}{e} \ln \frac{N_A N_D}{n_i^2}$$

The diffusion potential creates a potential barrier for further migration of free electrons from the n-type side to the p-type side and holes from the p-type side to the n-type side. That means diffusion potential prevents charge carriers to cross the junction. This region is highly resistive because of the depletion of free charge carriers in this region. The width of the depletion region depends on the applied bias voltage. The relation between the width of the depletion region and bias voltage can be represented by an equation called the Poisson equation.

$$W_D = \sqrt{\frac{2\epsilon}{e}(V_D - V)\left(\frac{1}{N_A + \frac{1}{N_D}}\right)}$$

Here, ε is the permittivity of the semiconductor and V is the biasing voltage. So, on an application of a forward-bias voltage, the width of the depletion region, i.e., P–N junction barrier decreases and ultimately disappears. Hence, in the absence of potential barriers across the junction in the forward-bias condition free electrons enter into the p-type region and holes enter into the n-type region, where they recombine and release a photon at each recombination. As a result, there will be a forward current flowing through the diode. The current through the P–N junction is expressed as

$$I = I_s\left(e^{\frac{eV}{kT}} - 1\right)$$

Here, voltage V is applied across the P–N junction and total current I, flows through the p–n junction. Is = reverse saturation current, e = charge of the electron, k is Boltzmann constant, and T is the temperature in Kelvin scale.

- **Ques: 220. Draw the V–I characteristics of the P–N junction diode.**

Solution:

When V is positive the junction is forward biased, and when V is negative, the junction is reverse biased. When V is negative and less than VTH, the current is minimal. But when V exceeds VTH, the current suddenly becomes very high. The voltage VTH is known as the threshold or cut-in voltage. For silicon diode VTH = 0.6 V. At a reverse voltage corresponding to point P, there is an abrupt increment in the reverse current. This portion of the characteristics is known as the breakdown region.

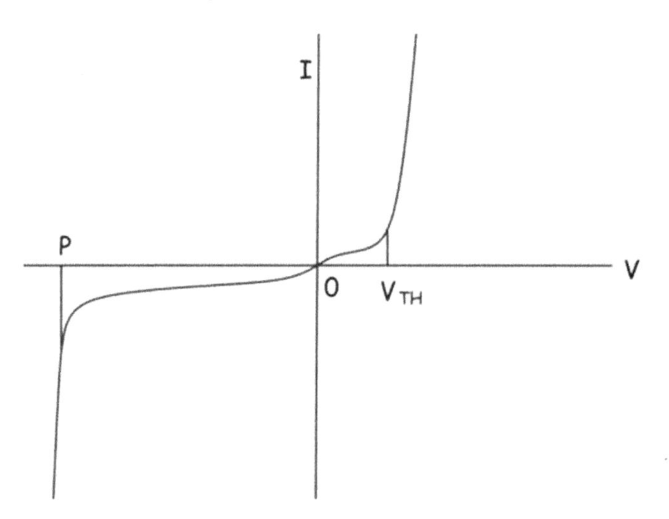

- **Ques: 221. Explain the concept of tunnel diode.**

Solution:

Tunnel diode is a type of semiconductor diode which is capable of very fast and in microwave frequency range. It was the quantum mechanical effect which is known as tunneling. It is ideal for fast oscillators and receivers for its negative slope characteristics. But it cannot be used in large ICs – that's why it's an application that is limited.

Tunnel diode is one of the most commonly used negative conductance devices. It is also known as Esaki diode after L. Esaki for his work on this effect. This diode is a two-terminal device. The concentration of dopants in both p and n region is very high. It is about 1024–1025 m-3 the P–N junction is also abrupt. For such reasons, the depletion layer width is very small. In the current–voltage characteristics of the tunnel diode, we can find a negative slope region when a forward bias is applied.

Quantum mechanical tunneling is responsible for the phenomenon and thus this device is named as tunnel diode. The doping is very high so at absolute zero temperature, the Fermi levels lie within the bias of the semiconductors. When no bias is applied any current can flow through the junction.

- **Ques: 222. Explain the characteristics of the tunnel diode.**

Solution:
When a reverse bias is applied the Fermi level of the p-side becomes higher than the Fermi level of the n-side. Hence, the tunneling of electrons from the balance band of the p-side to the conduction band of the n-side takes place. With the interments of the reverse bias, the tunnel current also increases.

When a forward bias is applied to the Fermi level of the n-side becomes higher than the Fermi level of the p-side, thus the tunneling of electrons from the n-side to the p-side takes place. The amount of the tunnel current is very large than the normal junction current. When the forward bias is increased, the tunnel current is increased up to a certain limit.

When the band edge of the n-side is the same as the Fermi level in the p-side, the tunnel current is maximum with the further increment in the forward bias the tunnel current decreases and we get the desired negative conduction region. When the forward bias is raised further, normal P–N junction current is obtained which is exponentially proportional to the applied voltage. The V–I characteristics of the tunnel diode are given in the following figure:

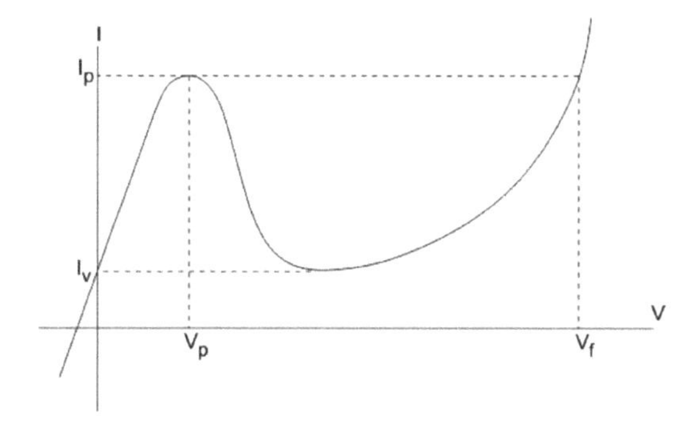

The negative resistance is used to achieve oscillation and often Ck+ function is of very high-frequency frequencies.

- **Ques: 222. Name any two applications of tunnel diode.**

Solution:
Tunnel diode is a type of semiconductor diode which is capable of very fast and in microwave frequency range. It was the quantum mechanical effect which is known as tunneling. It is ideal for fast oscillators and receivers for its negative slope characteristics. When the voltage is first applied current stars flowing through it. The current increases with the increase in voltage. Once the voltage rises high enough suddenly the current again starts increasing and tunnel diode stars behaving like a normal diode. Because of this unusual behavior, it can be used in several special applications as listed hereunder:

1. Oscillator circuits
Tunnel diodes can be used as high-frequency oscillators as the transition between the high electrical conductivity is very rapid. They can be used to create oscillation as high as 5 Gz. Even they are capable of creativity oscillation up to 100 GHz in an appropriate digital circuit.

2. Microwave circuits
Normal diode transistors do not perform well in microwave operation. So, for microwave generators and amplifiers tunnel diodes are used. In microwave waves and satellite communication equipment, they were used widely, but lately, their usage is decreasing rapidly, as transistors that operate in this frequency range are becoming available.

References

1. http://www.signoffsemi.com/sign-off-checks
2. Millman, J. (1967). *Electronic Devices and Circuits [by] Jacob Millman [and] Christos C. Halkias*. McGraw-Hill.
3. Bogart, T. F., Beasley, J. S., and Rico, G. (2004). *Electronic Devices and Circuits*. New Jersey: Pearson/Prentice Hall.
4. Gayakwad, R. A. and Gayakwad, R. A. (1988). *Op-amps and Linear Integrated Circuits* (Vol. 25). Englewood Cliffs: Prentice-Hall.
5. https://www.electrical4u.com/diode-working-principle-and-types-of-diode/

Annexure I: Digital Circuit IC Numbers

S. Nos.	Digital Logic	Parameters	IC/Board Nos.
1	Logic gates	Quad 2-input AND logic gate	7408
2		Quad 2-input OR logic gate	7432
3		NOT logic gate/hex inverter	7404
4		Quad 2-input NAND logic gate	7400
5		Quad 2-input NOR logic gate	7402
6		Quad 2-input Exclusive OR logic gate	7486
7		Quad 2-input Exclusive NOR logic gate	74266 (TTL) 4077 (CMOS)
8	Multiplexer	2:1 Multiplexer	74157
9		4:1 Multiplexer	74153
10		8:1 Multiplexer	74151
11		16:1 Multiplexer	74150
12	Demultiplexer	1:2 Demultiplexer	74LVC1G19
13		1:4 Demultiplexer	74139
14		1:8 Demultiplexer	74138
15		1:16 Demultiplexer	74154
16	Decoder	2: 4 Decoder	74155 (TTL)
17		3:8 Decoder	74137/74138
18		4:16 Decoder	74154
19		BCD to decimal decoder	7441
20		BCD to seven segment decoders	7446/7447
21	Encoder	8:3 Priority Encoder	74148
22		10:4 Priority Encoder	74147

S. Nos.	Digital Logic	Parameters	IC/Board Nos.
23	Digital Comparator	4-bit magnitude Comparator	7485
24		8-bit magnitude Comparator	74682
25	Flip-flop	SR Flip-flop	74279
26		JK Flip-flop	7470
27		JK Master Slave Flip-flop	7471
28		D Flip-flop	7474/7479
29		T Flip-flop	7473 short J & K
30	Shift register	8-bit Serial-In-Serial-Out (SISO) register	7491
31		8-bit Serial-In-Parallel-Out (SIPO) register	74164
32		16-bit Parallel-in-Serial-Out (PISO) register	74674
33		4-bit Parallel-in-Parallel-Out (PIPO) register	7495
34	ADC and DAC	16-bit A/D converter	ADS5482 (TI)
35		16- bit D/A converter	DAC8728 (TI)
36	Adder and Subtractor	2-bit Full Adder	7482
37		4-bit Full Adder	7483
38		4-bit Full Subtractor	74385
39	Counter	Up-down binary counter	74191
40		Up-down decade counter	74190
41		Modulo 10 counters	74416
42	Programmable Logic Devices (PLD)	Field Programmable Gate Array (FPGA)	SPARTAN 6 family, ARTIX 7 family
43		Complex Programmable Logic Device (CPLD)	ALTERA MAX 7000 series
44	Memories	16-bit RAM	7481/7484
45		64-bit RAM	7489
46		256-bit ROM	7488
47		512-bit ROM	74186 (open collector)

S. Nos.	Digital Logic	Parameters	IC/Board Nos.
48		256-bit PROM with open collector output	74188
49		2048-bit PROM with open collector output	74470
50		2048-bit PROM with three state output	74471
51		1024-bit PROM with three state output	74287

Annexure II: List of Keywords, System Tasks, and Compiler Directives Used in Verilog HDL

This list consists of keywords, system task, and compiler directives. All the keywords are defined in lowercase. The system tasks are *tasks* and functions that are used to generate input and output during simulation. The compiler directives are used to control the compilation of a *Verilog* description. The reference is IEEE std. 1364-2001, Verilog HDL.

Keywords		**System Tasks**	**Compiler Directives**
always	module	$bitstoreal	'accelerate
assign	task	$countdrivers	'autoexpand_vectornets
begin	library	$display	'celldefine
fork	time	$fclose	'default_nettype
case	table	$fdisplay	'define
casex	specify	$fmonitor	'define
casez	join	$fopen	'else
buf	end	$fstrobe	'elsif
bufif0	endcase	$fwrite	'endcelldefine
bufif1	endtable	$finish	'endif
rtran	endprimitive	$getpattern	'endprotect
rtranif0	endmodule	$history	'endprotected
rtranif1	endspecify	$incsave	'expand_vectornets
defparam	endtask	$input	'ifdef
deassign	pull0	$itor	'ifndef
include	pull1	$key	'include
integer	pullup	$list	'noaccelerate

Keywords		System Tasks	Compiler Directives
instance	pulldown	$log	'noexpand_vectornets
automatic	tri	$monitor	'noremove_gatenames
cell	tri0	$monitoroff	'nounconnected_drive
cmos	tri1	$monitoron	'protect
pmos	force	$nokey	'protected
nmos	forever	$stop	'remove_gatenames
and	real	$finish	'remove_netnames
or	reg	$write	'resetall
not	repeat	$rtoi	'timescale
nand	if	$readmemh	'unconnected_drive
nor	else	$readmemb	'undef
strong0	parameter	$hold	
strong1	primitive	$period	
supply0	wait	$skew	
supply1	wire	$timeformat	

Index

About the Authors

Dr. Cherry Bhargava is working as an associate professor and head, VLSI domain, School of Electrical and Electronics Engineering at Lovely Professional University, Punjab, India. She has more than 15 years of teaching and research experience. She is a Ph.D. (ECE), IKGPTU, M.Tech (VLSI Design & and CAD) Thapar University, and B.Tech (Electronics and Instrumentation) from Kurukshetra University. She is GATE qualified with an All India Rank of 428.

She has authored about 50 technical research papers in SCI, Scopus indexed quality journals, and national/international conferences. She has 12 books related to reliability, artificial intelligence, and digital electronics to her credit. She has registered 3 copyrights and filed 20 patents. She is recipient of various national and international awards for being outstanding faculty in engineering and excellent researcher. She is an active reviewer and editorial member of various prominent SCI and Scopus indexed journals. She is a lifetime member of the IET, IAENG, NSPE, IAOP, WASET, and reliability research group. Her area of expertise includes reliability of electronic systems, digital electronics, VLSI design, and artificial intelligence and related technologies.

Dr. Gaurav Mani Khanal is working as a post-doctoral researcher in DSP-VLSI lab, Department of Electronics Engineering, University of Rome Tor Vergata, Rome, Italy. He has earned his Ph.D. in Electronics Engineering (Memristor and Memristive System design and Fabrication) from University of Rome Tor Vergata, Rome (Italy). He has obtained an Advanced Master Degree in Wireless Systems and Related Technology from Politechnico Di Torino (Polytechnic University of Turin), Turin (Italy). He has completed his Master of Science (MS) in Microelectronics and System Design from Liverpool John Moores University, Liverpool (United Kingdom).

He possesses excellent knowledge of 2D material (esp. ZnO—Graphene and their composite) based low power semiconductor electronics devices (Theory and Fabrication), Memristor/Resistive switching, Device Modelling, Fabrication (Spin and Dip coating), and Characterization (XRD, SEM, and EIS). He has working experience with tools like Pspice, MATLAB, Model-Sim, Tanner, and Knowledge Verilog HDL for digital system design. He has hands on experience in application development with Arduino, Xilinx Spartan FPGA, and R-pi boards.